Geological Society of America
Memoir 186

Stratigraphy and Paleoenvironments of Late Quaternary Valley Fills on the Southern High Plains

Vance T. Holliday
Department of Geography
University of Wisconsin
Madison, Wisconsin 53706

with contributions by
Steven Bozarth
Scott A. Elias
Herbert Haas
Stephen A. Hall
Peter M. Jacobs
Raymond W. Neck
Barbara M. Winsborough

1995

Published by The Geological Society of America, Inc.
3300 Penrose Place, P.O. Box 9140, Boulder, Colorado 80301

Printed in U.S.A.

GSA Books Science Editor Richard A. Hoppin

Library of Congress Cataloging-in-Publication Data
Holliday, Vance T.
    Stratigraphy and paleoenvironments of Late Quaternary valley fills
on the southern High Plains / Vance T. Holliday ; with contributions
by Steven Bozarth ... [et al.].
      p.   cm. -- (Memoir / Geological Society of America ; 186)
    Includes bibliographical references and index.
    ISBN 0-8137-1186-X
    1. Geology, Stratigraphic -- Holocene. 2. Sedimentation and
deposition -- Texas. 3. Sedimentation and deposition -- New Mexico.
4. Paleoecology -- Texas. 5. Paleoecology -- New Mexico. 6. Valleys-
-Texas. 7. Valleys -- New Mexico.   I. Bozarth, Steven.   II. Title.
III. Series: Memoir (Geological Society of America) ; 186.
QE699.H695   1995
551.4'42'097648 -- dc20   95-34073
                                                        CIP

Contributors' addresses:

Steven Bozarth
Department of Geography, University of Kansas, Lawrence, Kansas 66045
Scott A. Elias
Institute of Arctic and Alpine Research, University of Colorado, Boulder, Colorado 80309-0450
Herbert Haas
Radiocarbon Laboratory, Desert Research Institute, Las Vegas, Nevada 89132
Stephen A. Hall
Department of Geography, University of Texas, Austin, Texas 78712-1098
Peter M. Jacobs
Department of Physics, Astronomy, and Geology, Valdosta State University, Valdosta, Georgia 31698
Raymond W. Neck
Houston Museum of Natural Science, One Hermann Circle Drive, Houston, Texas 77030
Barbara M. Winsborough
Winsborough Consulting, 5701 Bull Creek Road, Austin, Texas 78756

10 9 8 7 6 5 4 3 2 1

Dedication
to the men who led the way

E.H Sellards

E. H. Sellards
(1875–1961)

Grayson E. Meade   **Glen L. Evans**
(1912–1995)   **(1911–   )**

F. Earl Green
(1925–   )

I dedicate this volume to four gentlemen who were true pioneers in late Quaternary studies on the Llano Estacado. E. H. Sellards died many years before I began my research, but his work has long inspired me. Glen Evans, Grayson Meade, and Earl Green have had a more direct impact on my research, serving as models and mentors, and freely sharing observations and insights on their field work on the High Plains. Regrettably, Grayson Meade did not live to see this project completed. Sketch of E. H. Sellards reproduced with permission of the Association of American State Geologists.

Contents

GSA Data Repository 9541:*

Appendix A. Descriptions of Sites

Appendix B. Laboratory Methods and Data

Appendix C. Paleontological and Paleobotanical Methods

Appendix D. Historic Springs in the Draws of the Southern High Plains

*GSA Data Repository item 9541 is available on request from Document Secretary, Geological Society of America, P.O. Box 9140, Boulder, CO 80301.

Preface and acknowledgments

This project has its roots in my involvement with the Lubbock Lake archaeological project which I joined just after its beginning in 1973. I started on the project as an archaeologist, but over the years my interests grew to include the soils and stratigraphy of the site and now encompass the upper Cenozoic history of the Southern High Plains. My particular interest in the late Quaternary record of the draws culminated with the project I discuss in subsequent pages. Any research venture spanning so many years and so many miles, inevitably will involve a number of individuals and agencies. Whatever success I achieved in this work is due in large measure to the assistance of these people and organizations and it is with considerable pleasure that I acknowledge their support and thank them.

Funding for the 1988–1992 research was provided largely by the National Science Foundation, Surficial Processes Program (EAR-8803761). Additional funding was provided by the University of Wisconsin Graduate School in 1991 and 1992. Prior to 1988, the research was under the auspices of the Lubbock Lake Project, Eileen Johnson (Museum of Texas Tech University), Director, funded by the National Science Foundation, Moody Foundation (Galveston), West Texas Museum Association, and the Institute of Museum Research and the Museum, Texas Tech University.

Peter Jacobs, Ty Sabin, and Garry Running (graduate students in Geography, University of Wisconsin) were field assistants and I am indebted to them for their hard work, good humor, and insights. Jacobs also ran the laboratory with considerable diligence and efficiency for two years and produced Appendix B. Running and Sabin also helped with some lab work and Running generated Table 5 and Figure 10B. Additional lab assistance was provided by John Anderton, James Killian, Mark McCloskey, Drew Ross, and Jennifer Sibul.

I thank the authors whose paleontological and paleobotanical research enhanced the results of this study. R. M. Forester (U.S. Geological Survey, Denver) very kindly shared the results of his ostracode study for the later section, "Paleontology, Paleobotany, and Stable Isotopes".

I am indebted to Eileen Johnson for her many years of moral and financial support, particularly between 1988 and 1992, when she provided considerable logistical assistance. David Meltzer (Southern Methodist University) provided crucial logistical assistance on the southern Llano Estacado and was a stimulating research partner. Tom Gustavson (The University of Texas at Austin) arranged the loan of a Giddings rig for the 1989 season. The following graduate students helped trench at Gibson Ranch and the Midland Site: Paul Takac, Elizabeth Pintar, and Valentina Martinez (all of Southern Methodist University), and Andrea Freeman (The University of Arizona).

Many other colleagues provided additional assistance and advice: Vance Haynes (The

University of Arizona) on the Clovis site, upper Blackwater Draw, and regional paleoenvironments; John Montgomery and Joanne Dickenson (both of Eastern New Mexico University, Portales) for logistical assistance in and around the Clovis site; Eddie Guffee (Wayland Baptist University, Plainview) for access to the Plainview site and the Plainview Landfill, Glenn Goode (Texas Department of Transportation, Austin) for access to and data on the Quincy Street exposures; Dave Brown (Hicks & Co., Austin) for data we collected at the Lubbock Landfill, under contract with the City of Lubbock; Charles Frederick (Mariah Associates, Austin) for discussions of his work on lower Sulphur Springs Draw; Carolyn Spock, Darrell Creel, and Lynn Denton (The University of Texas at Austin) for providing information on the work of E. H. Sellards; Ken Honea (Northern Illinois University) for introducing me to the Marks Beach (Gibson Ranch) area; and Earl Green (Lubbock) for permission to publish his photographs of Blackwater Draw (Fig. 12). The manuscript benefitted from comments provided by S. C. Caran, C. R. Ferring, T. C. Gustavson, D. J. Meltzer, L. Nordt, C. C. Reeves, Jr., and two anonymous reviewers. Caran provided an exceptionally detailed review.

I thank the many landowners who allowed access to their land and who occasionally provided logistical support. All sites with the names of individuals (Appendix A) are named after landowners. Several provided support every year from 1988 through 1992: Clarence Scharbauer III (Midland site), and J. T. Gibson, Hollis Cain, and Charlotte Gibson Cain (Gibson Ranch).

Stephen Hall thanks Vaughn Bryant and Richard Holloway (Texas A&M University). Scott Elias thanks: Robert Gordon (U.S. National Museum of Natural History, Washington, D.C.) and Robert Anderson (Canadian Museum of Natural History, Ottawa) for assistance with species identification. Support for the fossil insect research at Lubbock Lake was provided by Texas Tech University and Texas Parks and Wildlife Department. Raymond Neck thanks Ron Ralph (Texas Parks and Wildlife Department) and Barbara Winsborough for providing some of the mullusc samples. Barbara Winsborough thanks Vance Haynes for providing the diatom samples from the Clovis site, and Michael K. Hein for preparing the photographic plates of diatoms.

The line drawings were prepared with support from the Cartography Laboratory of the Department of Geography at The University of Wisconsin–Madison. This chore was well executed by Gail Ambrosius and Joshua Hane, with assistance from QuingLing Wang, Sue Ryan, and Tim Ogg. Sharon Ruch (Department of Geography, University of Wisconsin) prepared all of the tables of field descriptions and laboratory data.

Finally, I thank my wife Diane for putting up with my summer field work.

Geological Society of America
Memoir 186
1995

Stratigraphy and Paleoenvironments of Late Quaternary Valley Fills on the Southern High Plains

ABSTRACT

The dry valleys or "draws" of the Southern High Plains (in northwestern Texas and eastern New Mexico), headwater tributaries of the Red, Brazos, and Colorado Rivers, contain late Quaternary sediments that accumulated over the past 12,000+ years. A few previous, scattered stratigraphic investigations of the draws strongly suggested synchroneity in late Quaternary depositional and soil-forming events and regionwide environmental changes. This volume reports on a systematic study conducted from 1988 to 1992 aimed at better documenting the late Quaternary geomorphic evolution and stratigraphic record of the draws, investigating their paleoenvironmental significance, and determining whether there were synchronous, regional, geomorphic, and soil-forming events in these dry valleys. The work focused on the past 12,000 years because most of the valley fill dates to this time, but older deposits occur locally and were investigated as well.

Most of the research was in Running Water, Blackwater, and Yellowhouse Draws (tributaries of the Brazos River), and Sulphur and Mustang Draws (tributaries of the Colorado River), with additional coring on McKenzie, Seminole, Monument, Monahans, and Midland Draws (all tributaries of the Colorado). Approximately 410 cores and exposures at 110 localities were studied. Samples collected from these sections underwent a variety of sedimentological and pedological analyses. Age control is provided by 53 new radiocarbon ages and scores of ages already available from several archaeological sites. Efforts also were made to recover pollen, phytoliths, molluscs, insects, ostracodes, and vertebrate faunal remains, and stable-carbon isotope trends were determined for four sites.

Several geomorphic processes and features exerted some influence on late Quaternary drainage development. Quaternary jointing and subsidence controlled drainage patterns around the margins of the Southern High Plains, particularly on the northern edge (Red River system). Major segments of most draws roughly parallel paleodrainages on the buried Tertiary erosion surface. Factors influencing the older drainage likely influenced development of the present drainage. Segments of Running Water, Blackwater, and Sulphur Draws also probably follow ancient drainageways that once connected the plains with the mountains to the west. Most of the draws intersect paleolake basins or extant lake bains, which may have exerted control on drainage development by directing water to paleotopographic lows or by overtopping and interconnection of basins.

The last phase of incision by the draws began after 20,000 yr B.P. but before 12,000 yr B.P. and aggradation began ca. 12,000 yr B.P. Valley fill predating this final incision is common locally and includes alluvial sand and gravel (stratum A) and lacustrine carbonate (stratum B). Eolian sheet sand with strong pedogenic modification (stratum C) accumulated on the uplands adjacent to some reaches during valley aggradation.

Holliday, V. T., 1995, Stratigraphy and Paleoenvironments of Late Quaternary Valley Fills on the Southern High Plains: Boulder, Colorado, Geological Society of America Memoir 186.

After ca. 12,000 yr B.P. the draws filled with a variety of sediments but produced a similar stratigraphic sequence among all of the drainages. Five principal lithostratigraphic units were identified: strata 1–5, oldest to youngest. Sandy and gravelly alluvium (stratum 1) is the oldest fill postdating final incision. There were several cycles of alluviation contemporaneous with or following the downcutting. The beginning of stratum 1 deposition is undated, but the top of stratum 1 ranges in age, with a few exceptions, from ca. 11,000 to ca. 9,500 yr B.P.

Stratum 2 contains beds of diatomaceous mud and noncalcareous or low-carbonate paludal mud conformably overlying stratum 1. Valley-margin facies of eolian and slopewash sands are common locally. A weakly developed soil formed at the top of stratum 2. These deposits are well known for containing extinct vertebrates and Paleoindian archaeological materials. Stratum 2 is quite rare, however. Of the >100 study localities only 12 yielded stratum 2. Beginning of stratum 2 deposition varied from ca. 11,000–ca. 10,000 yr B.P., and the end of deposition ranged from ca. 10,000–ca. 8,500 yr B.P.

Stratum 3 is a marl deposited by precipitation in marshes or shallow ponds along the valley axes. Locally the marl has a sandy, relatively low carbonate eolian facies along valley margins. A weakly developed soil formed at the top of stratum 3 (Yellowhouse soil). Most stratum 3 deposition occurred between ca. 10,000 and ca. 7,500 yr B.P., but both the beginning and end of deposition was time transgressive.

Stratum 4 is a thick (1-3 m), loamy to sandy, eolian layer. A moderately to strongly developed soil (Lubbock Lake soil: ochric or mollic A horizon over argillic and calcic Bt or Btk horizons) formed in stratum 4 and usually is the surface soil along the draws. Stratum 4 generally dates to ca. 7,500–4,500 yr B.P. The Lubbock Lake soil developed throughout the rest of the Holocene, except where buried by stratum 5. Stratum 5 includes localized accumulations of late Holocene paludal mud (beginning 3,900 yr B.P.) and slope wash and eolian sediment (beginning 3,000 yr B.P.).

The late Quaternary fill in the draws provides evidence of significant environmental change. From the latest Pleistocene to the early Holocene there was a hydrologic shift from flowing water (deposition of stratum 1) to standing water (deposition of strata 2 and/or 3), then almost complete disappearance of surface water and the accumulation of eolian sediment (stratum 4). Very broadly, the shifts in depositional environment were time transgressive (younger down draw). These environmental changes resulted from a decrease in effective regional precipitation from the late Pleistocene to the middle Holocene. In the late Pleistocene and early Holocene, local variability in the types and ages of the deposits was controlled by the presence or absence of springs and by time-transgressive decline in spring discharge. The early to middle Holocene eolian fill resulted from desiccation and wind deflation of the High Plains surface. By about 4,500 yr B.P., effective precipitation increased and vegetation became more dense, denoting establishment of the modern environment. There were brief climatic departures toward aridity in the late Holocene.

INTRODUCTION

For over a half century the largely dry valleys or "draws" of the Southern High Plains of northwest Texas and eastern New Mexico (Figs. 1, 2) have been known to contain a thick and varied stratigraphic record of sedimentation, soil formation, faunal and floral change, and human occupation spanning at least the past 12,000 years. This record is a key to understanding the late Quaternary paleoenvironments of the region, which is important because of the long history of human occupation of the area and because the High Plains is known to suffer from climatic extremes historically. Paleoenvironmental reconstructions may

help in understanding the more immediate climate and future climatic and other environmental changes.

Until 1988 the record of paleoenvironments and human prehistory in the draws was known largely from a handful of widely scattered archaeological sites (Fig. 2) and from a large interdisciplinary research project concerned with the late Pleistocene paleoecology of the region (Wendorf, 1961a; Wendorf and Hester, 1975). Most of the data on the draws came from just two sites, however: Clovis and Lubbock Lake (Fig. 2). These sites also are two of the better-known archaeological sites in North America due in part to the rich record of late Quaternary history preserved at the localities. The regional late Quaternary paleoenvironments

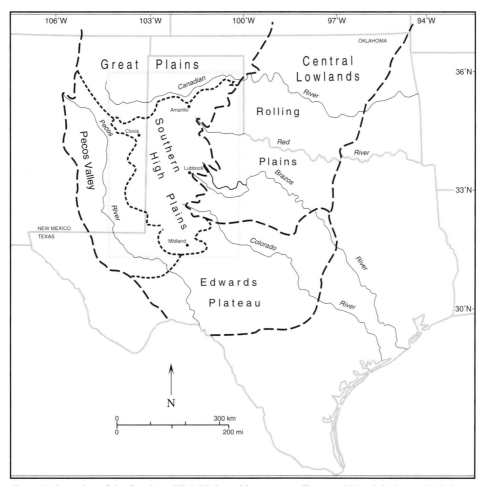

Figure 1. Location of the Southern High Plains with respect to Texas and New Mexico and relative to neighboring physiographic provinces and sections (base: Fenneman, 1931; Hunt, 1974). Boxed area is location of Figures 2, 3, and 4.

of the Southern High Plains were poorly known, however, in spite of over a half century of research. Studies at the archaeological sites were site specific and there were only a few such investigations scattered over a broad area. The late Pleistocene paleoecology project of the 1950s and 1960s was a landmark study in Quaternary interdisciplinary research, but depended largely on interpretations of pollen diagrams. With recent advances in palynology it appears that the early data were misinterpreted (Holliday, 1987; Hall and Valastro, 1995). A review of the results of all investigations before 1988 strongly suggested synchroneity in late Quaternary depositional and soil-forming events and regionwide environmental changes (Holliday 1985a, b, 1989a), but this reexamination presented only the broadest outline of the environmental history of the region, based on a few scattered sites.

In 1988 the author and collaborators began a systematic examination of late Quaternary sediments filling the draws that cross the Southern High Plains. This monograph presents the data and interpretations resulting from that study. The primary goal of the research was to document the late Quaternary strati-

graphic record preserved in the draws as an indicator of regional paleoenvironmental history. The field work focused on systematic coring of the draws, although the few available artificial exposures also were investigated. Approximately 410 cores and sections at over 100 localities in 10 draws were studied. Samples collected from these sections underwent a variety of sedimentological and pedological analyses. Age control is provided by several dozen new radiocarbon ages and scores of radiocarbon ages already available from several archaeological sites. Efforts were made to recover pollen and vertebrate and invertebrate faunal remains.

Environmental setting

The Southern High Plains or Llano Estacado ("stockaded plains" often translated as the "staked Plains") is a vast, level plateau covering approximately 130,000 km². It comprises the southernmost portion of the High Plains physiographic section (Fenneman, 1931; Hunt, 1974; Fig. 1). The plateau is defined on three sides by escarpments 50 to 200 m high. The western escarp-

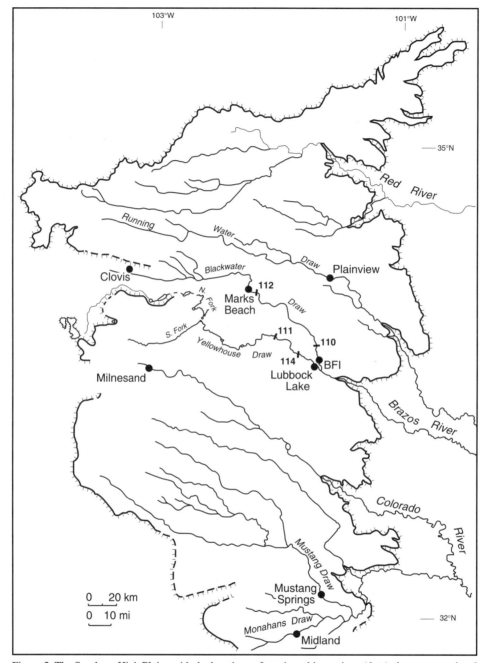

Figure 2. The Southern High Plains with the locations of stratigraphic sections (dots) along or associated with the draws reported prior to 1988. See Table 1 for references. Sites 110, 111, 112, and 114 are trenches reported by Stafford (1981).

ment separates the plateau from the Pecos River valley, and the northern escarpment separates the plateau from the Canadian River valley. Headward erosion by tributaries of the Red, Brazos, and Colorado Rivers formed the eastern escarpment, which separates the Southern High Plains from the Rolling (or Osage) Plains. The eastern escarpment provided the name Llano Estacado—the prominent topographic break took on the appearance of an immense stockade to Spanish explorers traveling west across the Rolling Plains (Bolton, 1990, p. 243). To the south, the surface of the Southern High Plains merges with the surface of the Edwards Plateau province of central Texas with only local topographic demarcation. The southern boundary is defined by the northernmost outcrops of Cretaceous Edwards Limestone, which characterizes the Edwards Plateau (Fenneman, 1931; Hunt, 1974; Fig. 1).

The climate of the Southern High Plains is continental and semiarid, classified as BScDw: steppe with dry winters, mainly mesothermal years (mean temperature of the coldest month is

0–18° C) with occasional microthermal years (mean temperature of the coldest month is below 0° C; Russell, 1945). There are relatively uniform gradients in precipitation and temperature across the region: precipitation generally increases from west to east and temperature usually increases from northwest to southeast. Mean annual precipitation across the region ranges from less than 35 cm in the southwest and northwest to over 50 cm in the northeast (Larkin and Bomar, 1983). Precipitation most commonly falls in the spring and summer. As is typical of semiarid, continental regions, the annual precipitation varies considerably from year to year. In Lubbock County, for example, the lowest total annual rainfall on record is 22 cm (1917) and the highest is 103 cm (1941; Blackstock, 1979). The region also experiences large temperature ranges and summer temperatures usually are quite high. The mean annual temperature for July and August generally ranges from 25–30° C (NOAA, 1982). Wind is an important climatic feature, blowing almost constantly across the open, flat landscape throughout the year. Average wind speeds range from 16 to 24 kph and speeds more than 80 kph are common (Lotspeich and Everhart, 1962; NOAA, 1982).

The natural vegetation of the Llano Estacado is a mixed-prairie grassland (Blair, 1950; Lotspeich and Everhart, 1962). The dominant native plant community is short-grass, which includes types of grama (*Bouteloua* sp.) and buffalo grass (*Buchloe dactyloides*). Trees are absent except along escarpment and reentrant canyons. The floristic composition varies somewhat from north to south due to changes in climate and soil texture. Native plant communities of the region occur in very few areas today, however, because most of the Southern High Plains is under cultivation. On a geologic time scale, the Llano Estacado was probably a grassland throughout the Holocene and probably ranged from a subhumid to semiarid savanna or sagebrush prairie to semiarid grassland in the Pleistocene as effective precipitation varied (Johnson, 1986, 1987a; Holliday, 1987, 1989a; Hall and Valastro, 1995).

The Southern High Plains is an almost featureless plain, ". . . the largest level plain of its kind in the United States" (NOAA, 1982, p. 3). There is a regional slope to the southeast with altitudes ranging from 1,700 m in the northwest to 750 m in the southeast. Slight topographic relief is provided by small basins, dunes, and dry valleys (Reeves, 1965, 1966, 1972; Wendorf, 1975a; Hawley et al., 1976; Walker, 1978; Holliday, 1985a; Fig. 3). There are about 25,000 small (< 5 km²) depressions dotting the landscape and containing seasonal lakes or "playas." There also are about 40 larger basins (tens of km²), also called playas, but are saline-lake basins or "salinas." The playa and salina basins contain the only available surface water on the Llano Estacado, although the water is seasonal and often brackish or saline. Lunettes (fringing dunes) are found adjacent to many of the playas and salinas, and several large sand dune fields are present along the western Llano Estacado (Fig. 3). The dry valleys or draws are northwest-southeast–trending tributaries of rivers on the Rolling Plains to the east. The draws are "elongate stream valleys having narrow drainage basins" (Gustavson and Finley,

1985, p. 33). Between the draws are expansive areas of the High Plains surface with no integrated drainage.

History of investigations

The first research on the draws dealt with the evolution of the Portales Valley, which includes upper Blackwater Draw (Baker, 1915; Theis, 1932; Bryan, 1938; Price, 1944). From 1933 to 1988, however, most of the work in the draws focused on the late Quaternary history within archaeological contexts (Table 1; Fig. 2). These studies began in 1933 with interdisciplinary archaeological work at the now-famous Clovis site, located in a small basin that once drained into Blackwater Draw proper. Work at Clovis was followed by the first archaeological studies at Lubbock Lake in Yellowhouse Draw in 1939. In 1945 the Plainview site in Running Water Draw was excavated; part of almost a quarter century of research by E. H. Sellards, G. L. Evans, and G. E. Meade (Texas Memorial Museum, The University of Texas at Austin) into the Quaternary geology and paleontology and Paleoindian archaeology of the Southern High Plains (Evans and Meade, 1945; Sellards, 1952; Sellards and Evans, 1960). Sellards, Evans, and Meade continued work at Clovis and Lubbock Lake and Sellards also worked at Midland in Monahans Draw. The initial and most substantive work at Midland was by F. Wendorf and colleagues, however.

In the late 1950s and early 1960s, Wendorf and others established the High Plains Paleoecology Project. This interdisciplinary research program, an outgrowth of the studies at Midland, primarily was aimed at reconstructing the late Pleistocene environment of the area but also addressed aspects of late Quaternary history including archaeology, geology, and Holocene paleoenvironments (e.g., Wendorf, 1961a; Green, 1962a, b, 1963; Hester, 1962; Hafsten, 1961, 1964; Kelley, 1974; Oldfield and Schoenwetter, 1964; Wendorf and Hester, 1975). This research program, which included work in the draws, playas, and dunes, was and remains one of the few large-scale, regional studies of late Quaternary history in North America that incorporated archaeology, stratigraphy, geochronology, paleobotany, and paleontology. The principal results of this research were environmental reconstructions indicating a sequence of late Quaternary "climatic intervals" and "pollen-analytical episodes" (Wendorf, 1961b, 1970, 1975b; Oldfield, 1975; Schoenwetter, 1975; Table 2). Results of some of this work were criticized (Holliday, 1987; Hall and Valastro, 1995), but the project provided a model of environmental change and a rich database for subsequent investigators.

In 1972 the Lubbock Lake Project was established, initially focusing on the record of human adaptation to late Quaternary environmental change as preserved at the site. The research now encompasses the Quaternary record throughout the Llano Estacado (e.g., Stafford, 1981; Holliday, 1985a, 1987, 1988a, 1989a, b; Johnson, 1986), including studies in draws at other archaeological and nonarchaeological sites (Table 1).

During the past several decades additional data on the draws became available. Most of the research focused on archaeologi-

Figure 3. Physiographic map of the Southern High Plains with locations of selected draws, dunes, saline playas, and rivers (base: 1:250,000 topographic maps, county soil surveys, and aerial photographs).

cal sites, including Clovis, Marks Beach (in Blackwater Draw), and Mustang Springs (in Mustang Draw). A few studies also dealt with drainage patterns (Reeves, 1970), drainage evolution (Hawley et al., 1976), and the stratigraphy and geomorphology of the Portales Valley (Reeves, 1972).

By 1988, archaeologic, stratigraphic, sedimentologic, pedologic, geomorphic, and geochronologic data from the draws were available based on over 50 years of work by many investigators at scattered localities. Some regional trends in stratigraphic successions, dominant depositional environments, and paleoenvironmental changes were apparent (Wendorf and Hester, 1975;

Stafford, 1981; Johnson, 1987b; Holliday, 1989a). The data indicated that (1) the late Quarternary stratigraphy in the draws of the central Llano Estacado is very similar, and (2) the paleoenvironments were similar at any one time over the region and environmental changes were roughly synchronous. Shortly before 11,000 yr B.P., Running Water, Blackwater, and Yellowhouse Draws held flowing water and the streams deposited sand and gravel. Lacustrine deposition in the form of diatomite and sapropelic mud began suddenly about 11,000 yr B.P. along some reaches, but streams continued to flow until about 10,000 yr B.P. along other reaches. By the earliest Holocene streams ceased to flow in

TABLE 1. RESEARCH AT SITES IN DRAWS PRIOR TO 1988*

Site	Nature of Research	Key References
BFI	Stratigraphy, soils	Holliday, 1983, 1985a.
Blackwater Draw (3 trenches)	Stratigraphy	Stafford, 1981.
Clovis (Blackwater Draw Locality 1)	Archaeology, stratigraphy, soils, paleontology, paleobotany	Antevs, 1935, 1949; Howard, 1935a, b; Stock and Bode, 1936; Cotter, 1937, 1938; Evans, 1951; Sellards, 1952; Green, 1962b, 1963, 1992; Haynes and Agogino, 1966; Hester, 1972, 1975; Stevens, 1973; Haynes, 1975, 1995; Holliday, 1985b; Boldurian, 1990; Stanford et al., 1990; Haynes et al., 1992.
Lubbock Lake	Archaeology, stratigraphy, soils, paleontology, paleobotany	Evans, 1949; Sellards, 1952; Green, 1962a; Black, 1974; Wheat, 1974; C. A. Johnson, 1974; Stafford, 1981; Holliday, 1985c, d, e, 1988a; Johnson, 1987b.
Marks Beach (Gibson Ranch)	Archaeology, stratigraphy, paleontology	Honea, 1980.
Midland (Scharbauer Ranch)	Archaeology, stratigraphy, paleontology	Wendorf et al., 1955; Sellards, 1955b; Wendorf and Krieger, 1959; Wendorf, 1975b.
Milnesand	Archaeology, paleontology	Sellards, 1955a.
Mustang Springs	Archaeology, stratigraphy	Meltzer and Collins, 1987; Meltzer, 1991.
Plainview	Archaeology, stratigraphy, soils, paleontology	Sellards et al., 1947; Sellards, 1952; Evans and Brand, 1956; Guffee, 1979; Holliday, 1985b, 1990b; Speer, 1990.
Yellowhouse Draw (2 trenches)	Stratigraphy	Stafford, 1981†

*Figure 2.
†Three trenches on Yellowhouse Draw are reported by Stafford, 1981, but one (Tr 109) was on the north side of Lubbock Lake and is considered part of the Lubbock Lake record.

all three draws and the valleys began to aggrade with organic-rich marsh sediments. These changes probably were the result of a significant reduction in runoff in the draws in the latest Pleistocene related to decreased effective precipitation and reduced spring discharge (Holliday, 1985c; Johnson, 1986). Eolian sedimentation began in the early Holocene and by the middle Holocene most of the marshes dried up and all the draws accumulated fine-grained eolian sediments, locally up to 3 m thick. These events indicate reduced effective precipitation relative to that of today and perhaps higher temperatures and increased seasonality (Holliday, 1985c, e, 1989a; Johnson and Holliday, 1986). Eolian deposition ended between 5,000 and 4,500 yr B.P. and there was little deposition or erosion during much of the late Holocene. As a result, the middle Holocene sediments often are strongly modified by pedogenesis. There was local deposition of eolian and paludal sediments and slope-wash deposits within the last several thousand years. The environments of the late Holocene probably were not dissimilar from those of today, with a few minor departures toward aridity (Holliday, 1985c; Johnson and Holliday, 1986). A generally similar geologic record is reported from Mustang Springs on the extreme southern part of the region (Meltzer and Collins, 1987; Meltzer, 1991) and significant middle Holocene eolian sediments are reported from several localities away from the draws (Gile, 1970; Holliday, 1989a).

This stratigraphic and paleoenvironmental record is important for several reasons. First, understanding paleoenvironmental changes and their effects on the landscape will aid researchers in predicting the impact of future environmental shifts in a region known historically for its sensitivity to such changes (Gustavson et al., 1990; Muhs and Maat, 1993). Second, documenting the late Quaternary stratigraphy, geochronology, and paleoenvironments of the Southern High Plains is important in analyzing and interpreting the record of human occupation in the region. Moreover, understanding the environmental history could be a key to understanding cultural evolution. The stratigraphy and paleoenvironmental history, however, was established on the basis of data from a handful of localities in valleys hundreds of kilometers long crossing an area of tens of thousands of square kilometers (Fig. 2). A better understanding of the late Quaternary history of the Southern High Plains requires a higher resolution database both temporally and spatially.

The 1988–1992 research

From 1988 to 1992 the author conducted a systematic investigation of the draws of the Southern High Plains, aimed at a more complete reconstruction of the late Quaternary paleoenvironments throughout the region. The work focused on the past 12,000 years because most of the valley fill dates to this time.

TABLE 2. CHRONOLOGY AND INTERPRETATIONS OF POLLEN-ANALYTICAL EPISODES*

Age (yr B.P.)	Pollen-Analytic Episode	Vegetation	Climate
	Post-Altithermal episodes[†]	Scattered pine?	Cooler?
	Sand Canyon postpluvial[†]	Grassland	Dry?
8,000	Unnamed subpluvial(s)(?)	Scattered pine?	Cooler?
9,000	Yellowhouse interval	Prairie	Drying
10,000	Lubbock subpluvial	Continuous pine woodland to mixed woodland and prairie	Cool-moist
	Scharbauer/White Lake[†] interval	Prairie	Dry?
11,000	Blackwater interval	Prairie and discontinuous pine and spruce(?) parkland	Cool-moist?
	Crane Lake interval[†]	Prairie	Dry?
12,000	Late Tahoka pluvial	Continuous pine and spruce woodland	Cool-wet
	Monahans/Vigo Park interval	Prairie and discontinuous pine and spruce woodland	Dry?
18,000	Early Tahoka pluvial	Continuous pine and spruce woodland	Cool-wet
ca. 27,000	Rich Lake interpluvial	Prairie or savanna	Cool-moist
ca. 32,000	Terry subpluvial	Prairie and discontinuous pine and spruce(?) woodland	Cool-wet
ca. 37,000	Arch Lake interpluvial	Prairie or savanna	Cool-moist
>37,000[§]	Brownfield oscillation	Prairie and discontinuous pine woodland (w/oak?)	Cool-wet

*By Oldfield, 1975; Schoenwetter, 1975; and Wendorf, 1961b, 1970, 1975b.
[†]Not recognized by Wendorf, 1975b.
[§]Given as <35,000 B.C. by Wendorf, 1975b, Table 14-1; apparently a typographic error.

Older deposits occur locally, however, and provide valuable clues to the development of the drainage system, which is an important component of regional landscape evolution.

The initial goal of the research was to determine if the stratigraphy is as regionally similar as the previous investigations suggest. Confirmation of such similarities would indicate: (1) that geomorphic responses to environmental changes were regionally similar and synchronous; (2) that past environments, like the modern ones, were regionally similar; and (3) that paleoenvironmental changes were synchronous throughout the area. These interpretations seem reasonable given the flat topography and relatively stable, uniform regional geology. Variations in the stratigraphy would suggest that local environmental factors such as topography or hydrology influenced the late Quaternary record. Beyond stratigraphy, the research focused on environmental reconstructions based largely on interpretations of depositional and pedogenic environments, with additional data provided by paleontology, paleobotany, and stable isotopes.

A variety of field and laboratory techniques were used in this study (Appendixes 1 and 2; and in GSA Data Repository 9541, Appendixes A, B, and C). The primary stratigraphic correlations and paleoenvironmental interpretations are based on lithologic and pedologic chracteristics of the sediment in the draws. The fill is composed of distinctive lithostratigraphic units, a characteristic noted by all investigators who worked in the draws since research began at Clovis in 1933. The distinct lithologies allow a first approximation of both stratigraphic correlations and depositional environments. The degree and nature of soil development also are useful indicators for dating and correlating some strata, for reconstructing landscape evolution in the draws during filling, and for making some paleoenvironmental reconstructions (Holliday, 1985b, c, d, e).

The field work focused on ten of more than a dozen named draws on the central and southern Llano Estacado (Fig. 4). All ten draws are part of the Brazos or Colorado systems, fully encompassing all previously reported draw localities (Fig. 2).

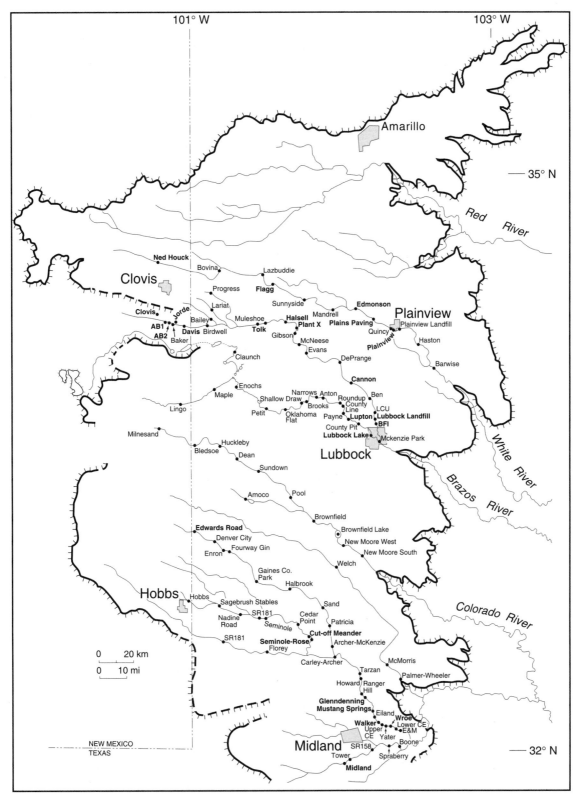

Figure 4. The Southern High Plains with locations of study sites along the draws of the Brazos and Colorado River systems. Site names in boldface are those referred to in other figure captions. Selected cities are also shown.

Three of the ten study draws are tributaries of the Brazos River: Running Water Draw, Blackwater Draw, and Yellowhouse Draw. The latter two meet in the city of Lubbock to form Yellowhouse Canyon. The work on Blackwater Draw also included limited coring on three longer tributaries of the draw (Bailey Draw, Lariat Draw, and Progress Draw). The other seven study draws are tributaries of the Colorado River: Sulphur Draw, McKenzie Draw, Seminole Draw, Monument Draw, Mustang Draw, Midland Draw, and Monahans Draw. Seminole and Monument Draws join to form Mustang Draw and McKenzie Draw joins Mustang below this confluence. These tributaries of Mustang are referred to as the Upper Mustang Draw system. Mustang Draw below the confluence with McKenzie is referred to as Lower Mustang Draw. Midland and Monahans Draws also join and feed into Mustang Draw. The only principal draw for which little data are available is Sulphur Springs Draw, a tributary of the Colorado (Fig. 3), although some information was gathered during the course of archaeological research along the middle reach of the draw (Johnson, 1994) and near the mouth (Frederick, 1993a, 1994) in the area of the Palmer-Wheeler locality (Fig. 4). Research in the Upper Mustang system was relatively limited due to problems of land access and the presence of extensive fields of oil and gas wells and pipelines that made coring and trenching dangerous.

Most of the data came from systematic coring of the draws. This was accomplished with a Giddings hydraulic soil-probe. Cores 5 cm in diameter and 120 cm per core-tube section were standard, with cores 7.5 cm in diameter taken for radiocarbon sampling. Hand-augering with a bucket-auger was done at a few localities. The few artificial exposures (quarries, landfills, and backhoe trenches) present along the draws also were studied. About half of the cores and sections were sampled for further description and possible laboratory characterization. Data are available from 110 sites (Fig. 4) including localities investigated prior to 1988. Over 400 cores, auger holes, trenches, and sections were investigated in addition to data, both published and unpublished, available from earlier studies. Descriptions of 33 selected cores and sections are presented in Appendix 1.

Approximately half of the samples were subjected to a variety of analyses for sedimentologic and pedologic characteristics, although the field characteristics and descriptions were the most informative kinds of data (e.g., Holliday, 1985b, c, d, e). Laboratory analyses included particle-size distribution (sand-silt-clay content), carbonate content, organic carbon content, bulk density, and clay mineralogy (GSA Data Repository 9541, Appendix B). Thin sections were prepared for some samples and were analyzed under a petrographic microscope.

Accurate dating of the draw stratigraphy is a key aspect of the study and 44 samples for radiocarbon dating were collected and submitted (Table 3; Appendix 2). Nine unpublished radiocarbon ages from other investigation of the valley fill also are available along with dozens of ages from several key sites such as Clovis and Lubbock Lake (Appendix 2). Ages were determined on organic-rich sediments and soil horizons. There are

problems in dating these materials (e.g., Campbell et al., 1967; Scharpenseel, 1971, 1979; Matthews, 1985), but with proper care in sampling, interpretation, and laboratory processing, they can provide reliable age control, especially if some time-diagnostic archaeological materials or dates on wood or charcoal are available for comparison (e.g., Holliday et al., 1983, 1985; Haas et al., 1986). For samples or horizons with several radiocarbon ages, the oldest age is considered the closest approximation of the true age (Matthews, 1980; Hammond et al., 1991). Contamination with dead carbon from ground water, precipitated in calcium carbonate, is the only known, common means of yielding falsely old ages in the region. Calcium carbonate is removed during processing of samples, however.

The radiocarbon ages are not tree-ring calibrated because (1) calibrations would confuse comparisons with other dated sequences from the region and surrounding areas, none of which are calibrated; (2) calibrations often require correction (especially the more recently published calibrations), rendering published, calibrated ages inaccurate (e.g., Stuiver and Pearson, 1992; Stuiver, 1993); and (3) many of the radiocarbon ages from the draws are in the range of only tentative calibrations (>10,000 yr B.P.; Becker, 1993; Stuiver, 1993). Calibration of only part of the sequence is not a useful exercise.

Plant and animal remains locally are abundant in the valley fill and consequently paleontology and paleobotany have long been important components of research in the draws (e.g., Stock and Bode, 1936; Patrick, 1938; Hafsten, 1961; Hohn and Hellerman, 1961; Oldfield and Schoenwetter, 1964; Lundelius, 1972; Slaughter, 1975; Johnson, 1987a, c; Pierce, 1987; Thompson, 1987). During the 1988-1992 studies, efforts were made to recover and analyze plant and animal remains (see later section, "Paleontology, Paleobotany, and Stable Isotopes"; GSA Data Repository 9541, Appendix C). Data on stable-carbon isotopes, which can indicate the composition of paleoflora, also are available from a few localities (see later section, "Paleontology, Paleobotany, and Stable Isotopes").

GEOLOGIC AND GEOMORPHIC BACKGROUND

Geologic setting

The oldest rocks exposed at the margins of the Southern High Plains are Permian mudstones of the Quartermaster Formation and Triassic mudstones and sandstones of the Dockum Group. Locally, Jurassic marine deposits unconformably overlie the Dockum Group. More extensive Cretaceous limestones, shales, and sandstones also unconformably overlie the Dockum Group. Uplift in the latest Cretaceous and Tertiary resulted in differential erosion of the Triassic, Jurassic, and Cretaceous strata (Harbour, 1975; Gustavson and Finley, 1985).

Extensive Cenozoic deposits overlay the Tertiary erosion surface and comprise most of the exposed sections and surficial deposits (Reeves, 1972; Hawley et al., 1976; Winkler, 1987; Gustavson and Winkler, 1988; Caran, 1991). Most of the Cenozoic deposits are eolian and alluvial sediment of the Ogallala Formation (Miocene-Pliocene; Fig. 5), ultimately derived from mountains to the west in New Mexico. There are several calcretes in the Ogallala, including a thick, highly resistant petrocalcic horizon in the upper Ogallala. This "caprock caliche" is a prominent ledge-forming unit near the top of the escarpments bordering the plateau. The calcrete is believed to be pedogenic and formed in upper Ogallala fine sand and silt.

The Blanco Formation (late Pliocene) is an extensive lacustrine layer of dolomite and some clastic sediment deposited in large basins cut into the Ogallala (Evans and Meade, 1945; Harbour, 1975; Hawley et al., 1976; Pierce, 1974; Holliday, 1988b; Caran, 1991; Fig. 5). A calcrete also formed at the top of the Blanco Formation. Many other, more localized, lacustrine deposits occur in the region, including the Tule Formation (early to middle Pleistocene) and the Double Lakes and Tahoka Formations (both late Pleistocene; Harbour, 1975; Reeves, 1976; Caran, 1991; Schultz, 1990a). Differentiating these Pliocene and Pleistocene lake sediments on lithological bases is very difficult (Holliday, 1995).

The Blackwater Draw Formation (Reeves, 1976; Fig. 5) is the major surficial deposit of the Southern High Plains, blanketing all older units. This formation consists of extensive, early to late Pleistocene eolian deposits and associated soils (Holliday, 1989b). The Pleistocene lake deposits are interbedded with these eolian sediments (Fig. 5).

The Ogallala Formation and superjacent units are important water-bearing deposits. This aquifer system, part of the Ogallala aquifer (or High Plains aquifer), which extends north to southern South Dakota, is one of the largest in North America and supports a major agricultural region (Weeks and Gutentag, 1988). On the Southern High Plains, most of the economically significant ground water is in the Ogallala Formation, but water in other Tertiary or Quaternary units hydraulically connected to the Ogallala is considered to be part of the Ogallala aquifer (Nativ, 1988; Weeks and Gutentag, 1988). The introduction of irrigation to the region early in the twentieth century and subsequent growth of the irrigated agriculture industry resulted in a dramatic decline of the water table (Cronin, 1969; Weeks and Gutentag, 1988). Besides its affect on agricultural practices, the drop in the water table also dried most surface water such as spring-fed streams and ponds (see GSA Data Repository 9541, Appendix D).

The late Quaternary (post–Blackwater Draw Formation) stratigraphic record of the Southern High Plains is found in the lake basins, dunes, and draws of the High Plains surface (Reeves, 1972; Harbour, 1975; Holliday, 1985a, 1995; Fig. 5). The playa and salina basins are cut or collapsed into the Blackwater Draw Formation or older units and contain late Pleistocene and Holocene lacustrine and paludal sediments. The lunettes and dune fields rest on top of the Blackwater Draw Formation. The lunettes contain late Pleistocene and Holocene eolian sediments and the dune fields consist primarily of Holocene sediments.

The draws, the subject of this monograph, are relatively broad, shallow valleys inset into the Blackwater Draw Formation and locally into older deposits (Tables 4, 5). Within these now-dry tributaries are late Quaternary lacustrine, paludal, alluvial, and eolian deposits divided into Older Valley Fill and Younger Valley Fill (Fig. 6). The older deposits, discussed below, are scattered remnants of late Pleistocene eolian, lacustrine, and alluvial sediments that were incised by final downcutting of the draws or isolated on uplands adjacent to the draws. The younger deposits, discussed in the following section, "Stratigraphy of the Valley Fills", are the largely continuous latest Pleistocene and Holocene sediments that filled the draws following the final phase of downcutting.

Origins and early history of the draws

The origin of the present draw systems is associated with several geologic and geomorphic processes and features, each of which probably exerted some influence on late Quaternary drainage development: Quaternary faulting and subsidence, paleodrainage systems, paleolake basins, extant lake basins, and springs. As illustrated in the following discussion, the influence of each process and feature varies in degree geographically and temporally.

Fracturing or faulting was proposed as a control on some elements of the drainage pattern (Finch and Wright, 1970; Reeves, 1970). In response, Gustavson and Finley (1985, p. 33) comment that structure-contour maps on Permian formations show no evidence of deep-seated tectonic faulting. Fracturing and infiltration of ground water, with resultant dissolution of bedded Permian salts, may exert some influence on drainage development, however. Gustavson and Finley (1985) show that orientation of the Red River draws (Fig. 3) probably is directly controlled by dissolution of salt in the Paleozoic bedrock and subsidence of the overlying beds. South of the Red River system salt dissolution probably played a lesser role in development of the Quaternary drainages. Gustavson and Finley (1985) see relatively little evidence for the removal of salt in the Paleozoic rock under the Llano Estacado south of the Red River system. The rectilinear pattern of some draws (most notably the "dog leg" on lower Sulphur Springs Draw and the straight, parallel reaches of upper Running Water and upper Frio Draws, Fig. 3), however, indicates possible structural control on drainage development.

Gustavson and Finley (1985, p. 21) suggest that the parallelism and superpositioning of modern drainages over Tertiary drainages indicate that factors influencing the older drainage also influenced development of the present channel network. In the area crossed by draws of the Red and Brazos River systems the surface drainage generally coincides with that existing on the eroded surface of the Paleozoic and Mesozoic bedrock exposed just prior to deposition of the Ogallala Formation (Gustavson et al., 1980, fig. 8; Gazdar, 1981; Gustavson and Finley, 1985, p. 21; Fig. 7). "Major segments of both drainage systems are roughly parallel, and modern streams tend to occur near the paleostreams in some areas" (Gustavson and Finley, 1985, p. 21). Parallelism between the surface drainage and pre-Ogallala drainage also is

TABLE 3. GEOCHRONOLOGICAL SUMMARY FOR THE VALLEY FILL*

Stratum	Draw	Site†	Dating§ (yr. B.P.)
5	Blackwater	Evans	340 (5s1 top)
		Clovis	<1,000 (5s)
	Lower Mustang	Mustang Spring	>2,000-100 (5m)
	Running Water	Flagg	<770 (5s)
		Quincy St.	2,600 (5m, burial by 5s)
		Plainview	3,200 (5m base)
			<3,900 (5s1); <2,100 (5s2)
	Sulphur	Sundown	<2,600 (5s)
	Sulphur Springs	Lower reach	420 (5m top)**
			3,000 (5m base)**
	Yellowhouse	Lubbock Lake	800-100 (5s and 5g); 3,900-100 (5m)
4	Blackwater	Lubbock Landfill	3,100 (4s top; burial by 5s)
		Clovis	5,000 (4s base)
		Lubbock Landfill	6,500 (3si top; burial by 4s)
		BFI	<8,100 (4s)
		Halsell	<8,500 (4s)
		Bailey	<9,000 (4s)
	Lower Mustang	Wroe	5,700 (3c top; burial by 4s)
		Mustang Spring	6,800 (4s base)
		Walker	<8,700 (4s)
		Glendenning	<9,500 (4s)
	Running Water	Flagg	770 (4s top; burial by 5s)
			<8,600 (4s base)
		Plainview	3,900 (4s top; burial by 5s1)
			2,100 (4s top; burial by 5s2)
			<6,800 (4s)
		Houck	<9,800 (4s)
	Sulphur	Sundown	2,600 (4s top; burial by 5s)
		Brownfield	9,700 (4s base)
	Yellowhouse	Lubbock Lake	1,200-800 (4s top; burial by 5s and 5 g)
			5,500-4,500 (4s)
		Enochs	8,600 (4s base)
3	Blackwater	Evans	3,500 (3c top; burial by 5s)
		Anderson Basin #1	7,300 (3s middle)
		Cannon	7,900 (3c top; burial by 4s)
		BFI	8,100 (3c top; burial by 4s)
			5,100 (3s top; burial by 4s)
		Halsell	8,500 (3c top; burial by 4s)
		Clovis	9,000-8,000 (3s)
		Bailey	9,600 (3c base); 9,000 (3c top; burial by 4s)
		Lubbock Landfill	<10,000-8,800 (3c); <8,800->6,500 (3si)
	Lower Mustang	Mustang Springs	6,900 (3c base)
		Walker	8,700 (3c top; burial by 4s)
		Glendenning	9,500 (3c top; burial by 4s)
		Wroe	8,300 (3c) - 5,700 (3c top; burial by 4s)
	Midland	Boone	9,500 (3c middle)
	Running Water	Plainview	8,900 (3c) - 6,800 (3s top; burial by 4s)
		Plains Paving	<9,300
		Edmonson	9,800 (3c base)
		Houck	9,800 (3c top; burial by 4s)
	Sulphur	Brownfield	9,700 (3c top; burial by 4s)
	Sulphur Springs	Lower reach	6,300 (3c)**
	Yellowhouse	Lubbock Lake	6,300-5,500 (3c)
		Enochs	>8,600 (3c)
2	Blackwater	Gibson	9,700 (2d)
		Tolk	10,100 (2m)
		Anderson Basin #1	10,700 (2m)
		Davis	10,900 (2m)

TABLE 3. GEOCHRONOLOGICAL SUMMARY FOR THE VALLEY FILL* (continued)

Stratum	Draw	Site†	Dating§ (yr. B.P.)
2 (continued)		Clovis	10,800-10,000 (2d); 10,000-8,500 (2s)
		Lubbock Landfill	>10,500-10,200 (upper 2m)
		Progress	11,300-9,900 (2m)
	Lower Mustang	Mustang Springs	10,100 (2d base) - 8,100 (2m top)
		Wroe	10,100 (2m)
	Running Water	Plainview	8,900 (2m)
		Plains Paving	9,300 (2m)
		Flagg	9,400 (2m) - 8,600 (2d)
		Edmonson	9,800 (2m top and 3c base)
	Sulphur Springs	Lower reach	8,700 (2m top)**
	Yellowhouse	Lubbock Lake	11,000-10,000 (2d); 10,000-8,500 (2m); >9,000-<7,800 (2s); 8,500-6,300 (soil formation)
1	Blackwater	Bailey	>9,600 (1 top)
		Gibson	>9,700 (1 top)
		Tolk	>10,100 (1 top)
		Anderson Basin #1	>10,700 (1 top)
		Davis	>10,900 (1 top)
		Clovis	11,500-11,000 (upper 1)
		Lubbock Landfill	>10,500 (1 top)
		Progress	>11,300 (1 top)
	Lower Mustang	Mustang Springs	>10,100 (1 top)
		Wroe	>10,100 (1 top)
	Running Water	Plainview	<12,000->8,900 (1 top)
		Plains Paving	>9,300 (1 top)
		Flagg	>9,400 (1 top)
		Edmonson	>9,800 (1 top)
		Houck	>9,800 (1 top)
	Yellowhouse	Lubbock Lake	11,100 (1 top)

*See Appendix 2 for full information on radiocarbon ages.
†Located on Figure 4.
§Ages given are approximations based on rounding of radiocarbon means (see Appendix 2 and Table A2.1); parenthetical notations refer to strata.
**Radiocarbon ages for lower Sulphur Springs Draw are presented and discussed by Quigg et al., 1994.

apparent in the southern half of the Llano Estacado (the Mustang Draw system of the Colorado drainage; Cronin, 1969, sheet 1; Walker, 1978, fig. 17; Gazdar, 1981; Fig. 7). The draws of the Brazos and Colorado systems were, therefore, likely influenced by antecedent (pre-Ogallala) topography and possibly some late Cenozoic dissolution and subsidence, whereas the draws of the Red River system are more directly affected by relatively recent dissolution and subsidence.

The drainage present on the post-Ogallala surface probably maintained its configuration through the Pleistocene because deposition of the Blackwater Draw Formation was a slow, accretionary process. The only field data to support this hypothesis comes from the Lubbock Landfill (discussed below), but a similar scenario is emerging from field investigations of playa basins (V. T. Holliday, unpublished data; T. C. Gustavson and S. Hovorka, personal communication, 1994).

Several large, ancient lake basins are known on the Llano Estacado (Fig. 7). Most are found in the lower reaches of the modern draw systems and several are associated with the pre-Ogallala drainage. The processes that formed the basins (e.g., subsidence) and the presence of the topographic lows may have influenced development of the respective drainages (Gustavson and Finley, 1985). Along the Brazos and Colorado systems these basins include the Blanco basin along lower reaches of Running Water Draw, Blackwater Draw, and Yellowhouse Draw (Gustavson and Finley, 1985), the Anton basin in the Yellowhouse drainage (Bartolino, 1991; Reeves, 1990), the Patricia basin in lower McKenzie Draw and its confluence with Mustang Draw, and the Lomax basin in Sulphur Springs Draw (Frye and Leonard, 1968). The Blanco basin is an ancient topographic feature containing the type lacustrine deposits for the Blanco Formation and the type fauna of the Blancan Land Mammal age (Meade, 1945; Schultz, 1990b; Holliday, 1988b). The extent of the basin is not known, but the region of Blanco outcrop, which extends from the Blanco type area westward to the Lubbock area overlies a large pre-Ogallala paleotopographic low with evidence

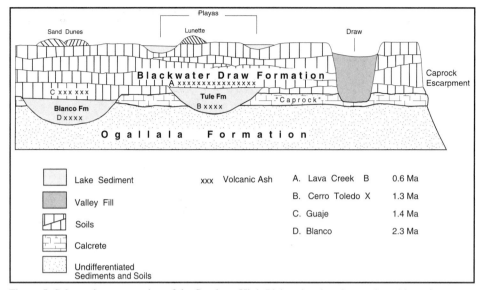

Figure 5. Schematic cross section of the Southern High Plains showing the stratigraphic and geomorphic relationships of most upper Cenozoic units. Only selected Pleistocene lacustrine formations are shown. No vertical or horizontal scale is implied. Based on data from Hawley et al. (1976), Holliday (1985a, 1995), Caran (1991), and Gustavson et al. (1990).

of salt dissolution (Cronin, 1969, sheet 1; Gustavson and Finley, 1985, p. 34; Fig. 7).

The Anton basin contains Pleistocene and probably upper Tertiary lake sediments. A striking topographic characteristic of this basin is escarpments composed of the Ogallala Caprock facing toward the basin on its west and south sides. Yellowhouse Draw enters the basin from the northwest, incising the Caprock escarpment and forming a small canyon, here termed the "Narrows" (Fig. 4). The draw crosses the north side of the basin and then swings away to the northeast. The draw then curves to the southeast and enters a playa basin. According to Gustavson and Finley (1985, figs. 8, 9, 10) the Anton basin is located over the northwestern end of a linear pre-Ogallala depression. The playa basin intersected by the Yellowhouse down draw from the Anton basin is coincident with the southeastern end of the pre-Ogallala depression. Reeves (1990) proposed that at one time the ancient Yellowhouse flowed through the Anton basin and on to the southeast. If so, this route would be quite old as new data from the Lupton site, down draw (Fig. 4; discussed below), shows that it has been in its present position since the late Pleistocene.

The Patricia and Lomax basins contain late Pleistocene sediments and faunas. The Lomax basin is the largest late Pleistocene lake basin reported for the High Plains (Frye and Leonard, 1964, 1968), although the ancient lake itself probably occupied only a small portion of the basin (Frederick, 1993a, 1994). Lake Lomax was at the mouth of Sulphur Springs Draw in a large topographic basin on the edge of the High Plains escarpment (Fig. 7). The Patricia basin is a broad, shallow topographic basin with late Pleistocene sediments (Eifler and Reeves, 1976) in lower McKenzie Draw and includes the confluence with Mustang Draw (Fig. 7). In the confluence area the two draws are

incised into the lake sediments. The basin fill locally must be older than the Holocene-age valley fill. The lower reaches of both Sulphur Springs and McKenzie Draws and their respective lake basins coincide with paleotopographic lows on the pre-Ogallala surface. The Lomax and Patricia basins probably were fed by the modern draws based on the age of the basin fills and, therefore, the lower reaches of the draws were probably affected by events in the respective basins.

Several draws also connect or are related to existing lake basins (Fig. 3). This led to some speculation that development of these drainages is related to interconnection between lakes (Reeves, 1966). Connected lake basins are particularly characteristic of Yellowhouse Draw (Fig. 8). The North and South Forks of Yellowhouse connect several salinas. The middle and lower Yellowhouse connects and is incised into a series of smaller basins similar to the small playa basins that dot the surface of the High Plains. There is no evidence for an antecedent drainage system on the pre-Ogallala surface under most of Yellowhouse Draw. The basins may have been, therefore, a dominant factor in the evolution of the Yellowhouse drainage as basins were interconnected by water overtopping divides under wetter ("pluvial") climates (J. W. Hawley, personal communication, 1980; G. L. Evans, notes on file, Museum of Texas Tech University, n.d.).

At least some segments of nine draws cut through Pleistocene or older lacustrine sediments not associated with the large, ancient lake basins discussed above (Table 4; Fig. 7), further suggesting a link between drainage location and paleolake basins. Several of these sites are discussed in the following section. Characteristics of the lacustrine deposits and their settings include: a relatively restricted areal extent; no obvious topographic indication of the paleobasins; and no correlation between

TABLE 4. RELATIONSHIP OF DRAWS TO BEDROCK*

Draw†	Bedrock Characteristics§
Blackwater	Reaches in New Mexico inset against Older Valley Fill of ancient Portales Valley and against Blackwater Draw Formation. In Texas draw cut through the Blackwater Draw Formation; from LCU site to confluence with Yellowhouse Draw, cut into Ogallala Formation; cut into or through paleolake carbonates at Anderson Basin #2 (upper Pleistocene), Bailey Draw, Clovis, Davis, Gibson, LCU, Lubbock Landfill (Blanco Formation), Owen Ben (late Pleistocene).
Lower Mustang	Cut through Blackwater Draw Formation and into the Ogallala Caprock calcrete; cut into Cretaceous limestone near confluence with McKenzie Draw (Eifler and Reeves, 1976); cut into or through paleolake carbonate at Carley-Archer, Tarzan, Curtis Erwin (upper and lower), Evans and Meade (late Pleistocene).
McKenzie	Cut through Blackwater Draw Formation and into Ogallala Caprock calcrete in most reaches; cut into or through paleolake carbonate at headwaters area (New Mexico), Four-way Gin, Patricia, Archer-McKenzie.
Running Water	Most reaches cut through Blackwater Draw Formation and into or through the Ogallala Caprock calcrete; cut into or through paleolake carbonates at Bovina, Lazbuddie, Mandrell, Plains Paving, Plainview Landfill.
Seminole	Cut through Blackwater Draw Formation and into Ogallala Caprock calcrete in most reaches; cut into or through paleolake carbonate at Nadine Road, confluence with Wardswell Draw, Cut-off Meander, Seminole-Rose (late Pleistocene).
Sulphur	Cut through Blackwater Draw Formation and into Ogallala Caprock calcrete from Huckleby to Brownfield; cut into or through paleolake carbonate at Bledsoe, Pool, Brownfield Lake (late Pleistocene).
Yellowhouse	Cut through Blackwater Draw Formation and into Ogallala Formation along both forks and below their confluence to the Narrows; small canyon cut into the Ogallala Caprock calcrete at Narrows where draw enters Anton Basin; below Narrows, cut through Blackwater Draw Formation; cut into or through paleolake carbonates at Anton Basin (late Pleistocene), County Line, Roundup, Payne, Lupton (late Pleistocene), County Caliche Pit (Blanco Formation), Lubbock Lake (Blanco Formation).

*Lithostratigraphy or chronostratigraphy of bedrock indicated if known.
†See Figure 3 for locations.
§See Figure 4 for locations.

these deposits and pre-Ogallala basins. The correlation between the location of the draws and these isolated deposits may be more apparent than real; buried lake sediments may be common throughout or below the Blackwater Draw Formation and therefore are fortuitously exposed by the draws.

Drainage development also may be linked to spring activity. Historic data show that Running Water and Blackwater Draws, the two largest, had the highest concentration of springs among all of the draws of the Brazos and Colorado systems (Brune, 1981; GSA Data Repository 9541, Appendix D). The data from coring also suggest that springs were most common along these two draws throughout the late Quaternary (discussed more fully in the later section, "Summary, Discussion and Conclusions"). Likewise, on Mustang Draw historic and ancient spring activity was significantly higher at and below Mustang Springs than it was above the springs, coinciding with an increase in slope and general deepening and widening of the draw down stream. Drainage characteristics, therefore, may be linked to spring activity, but the data are equivocal. Higher discharges due to spring flow could result in the development of wider, deeper, and perhaps steeper valleys. Alternatively, the development of deeper valleys intersecting the water table could lead to increased spring activity.

The Older Valley Fill and development of individual drainages

In the course of investigating the late Quaternary record of the draws, three stratigraphic units pertaining directly or indirectly to the evolution of the valleys were identified and grouped as the Older Valley Fill: strata A, B, and C. Stratum A is composed of fine sand and locally includes some gravel. A well-expressed, clayey A horizon is present locally at the top of stratum A (Fig. 9A). Stratum A rests unconformably on the Blackwater Draw Formation and usually is overlain conformably by stratum B. The A horizon formed in the top of the Blackwater Draw Formation where stratum A is missing, and stratum B overlies the Blackwater Draw directly. Stratum B is a massive lacustrine carbonate not otherwise correlative with the Blanco, Tule, or Tahoka Formations. A dark gray, clayey A horizon is present locally at the top of stratum B (Fig. 9B) and lenses of calcareous sand sometimes are found within the carbonate. Stratum C occurs as a sheet of loamy to sandy eolian sediment on the uplands, usually resting unconformably on stratum B. An A-Bt soil profile with 7.5YR hues, thin clay films, and weak structural expression developed in stratum C, but is not as well expressed as those in the Blackwater Draw Formation. Exposures of the older strata at the Seminole-Rose road cut and at the Clovis site (Figs. 4, 9) provide good reference sections for the Older Valley Fill.

Most speculation on development of individual draws focuses on upper Blackwater Draw. The upper 40 km (25 mi) of the draw in New Mexico is in a broad valley called the Portales Valley, which becomes part of a westward-draining reentrant on

TABLE 5. MORPHOMETRIC STATISTICS FOR THE STUDY DRAWS*

Draw[†]	N[§]	Width Minimum (m)	Width Maximum (m)	Width Mean and Std. Dev.	Depth Minimum (m)	Depth Maximum (m)	Depth Mean and Std. Dev.	Sinuosity	Slope (m/km)	Length (km)
Brazos River System										
Bw	15	213	1,874	775 ± 515	2	27	11 ± 7	1.13 ± 0.05	1.3	233
Rw	10	640	2,103	986 ± 446	4	27	16 ± 7	1.17 ± 0.04	1.7	253
Yh**	15	183	823	462 ± 197	3	13	7 ± 3	1.10 ± 0.02	1.9	175
Colorado River System										
McK	7	152	427	253 ± 101	5	9	6 ± 2	1.09 ± 0.03	1.9	151
Mn	4	122	335	208 ± 89	2	6	4 ± 2	1.04 ± 0.02	1.6	73
Mu‡	8	152	559	304 ± 122	4	15	8 ± 4	1.28 ± 0.04	1.0	131
Se	8	305	945	482 ± 219	6	15	10 ± 3	1.20 ± 0.10	1.7	119
Su	8	121	427	297 ± 111	3	12	7 ± 2	1.18 ± 0.04	1.8	159

*Monument and Midland not included because of few study sites.

†Bw = Blackwater, McK = McKenzie, Mn = Monahans, Mu = Mustang, Rw = Running Water, Se = Seminole, Su = Sulphur, Yh = Yellowhouse. See Figure 3 for locations.

§N = number of reaches measured.

**Measured sites are only along the South Fork and Lower Yellowhouse (i.e., below the confluence of the north and south forks).

‡Measured sites on Lower Mustang only (i.e., below confluence of Mustang and Seminole Draws).

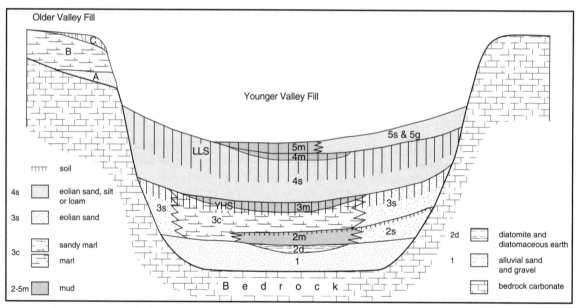

Figure 6. Schematic illustration of stratigraphic relationships within the Older Valley Fill and within the Younger Valley Fill (LLS = Lubbock Lake soil; YHS = Yellowhouse soil). Strata 3m, 4m, and 5g are localized and not shown on Figure 19. No vertical or horizontal scale is implied. Key is for all stratigraphic cross sections.

the west side of the Llano Estacado (Figs. 3, 10A). The Portales Valley probably is a segment of the ancient Brazos River system that once included the upper Pecos drainage and flowed across the High Plains prior to diversion of the drainage to the south as the modern Pecos system developed (Baker, 1915; Theis, 1932; Fiedler and Nye, 1933; Finch and Wright, 1970; Kelley, 1972; Reeves, 1972; Gustavson and Finley, 1985).

Interpretations differ on the extent of the Portales Valley (Theis, 1932; Reeves, 1972; Price, 1944; Fig. 10A). These dif-

ferences are due to the indistinct topographic boundaries of the valley except for escarpments south and southeast of Portales, New Mexico. Topographic profiles through the area (Fig. 10B) illustrate the presence of a broad depression 30–35 km wide (north to south) in and east of the area of Portales. This broad basin is here referred to as the Portales Valley and the deepest part of the basin is the Inner Portales Valley (Fig. 10B). Blackwater Draw is within the Portales Valley, north of and roughly parallel to the Inner Portales Valley (Fig. 10A, B). The topo-

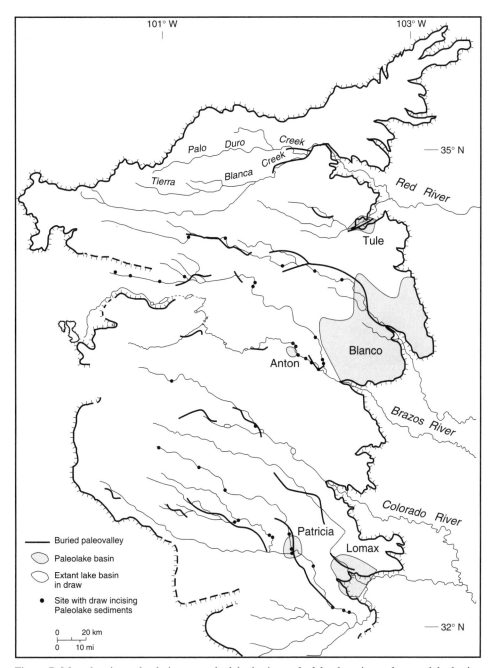

Figure 7. Map showing paleodrainages, paleolake basins, paleolake deposits, and extant lake basins associated with draws. Paleodrainages are those segments of valleys on the pre-Ogalalla surface that roughly parallel the modern drainage (based on: Cronin, 1969, sheet 1; Gustavson et al., 1980, fig. 8; Gazdar, 1981, figs. 11, 17; Gustavson and Finley, 1985, fig. 10). Paleobasins are the major structural lake basins coincident with the draws. All other sites are those with outcrops of lake sediment incised by the draws (see Table 4 for listing of sites).

graphic expression of the Portales Valley becomes difficult to fol- low east of the state line because the paths of Blackwater Draw and the Inner Portales Valley diverge (discussed further below) and the divide between the drainages becomes indistinguishable from the High Plains surface. Price (1944, fig. 33), however, extends the valley eastward into Texas almost 100 km (60 mi).

Proposed routes of the ancestral Brazos River include the Blackwater/Yellowhouse Canyon drainage, the Running Water/ White River drainage, or both (Gustavson and Finley, 1985, p. 37). Reeves (1972, p. 113) presents the most substantive dis- cussion, based on contours on the Quaternary fill in the valley and on the base of the Ogallala Formation to the east. He pro- poses "that the ancient Portales [ancestral Brazos] River flowed northwest of the Coyote Lake area where it joined a southeast-

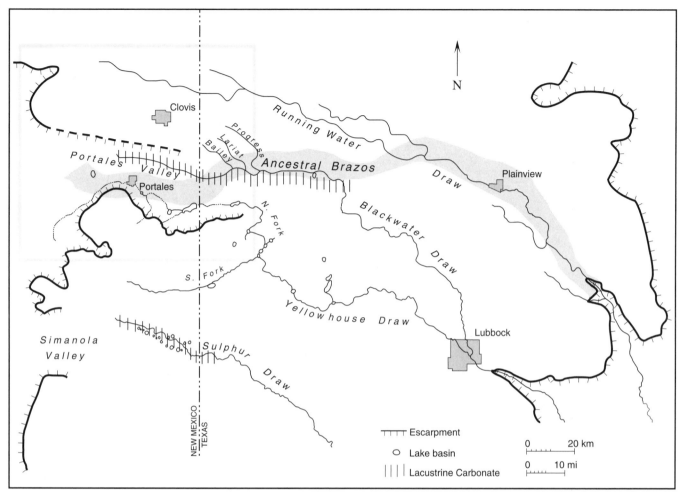

Figure 8. Map showing the relationship of draws to reentrant valleys in the western escarpment of the High Plains, the route of the ancestral Brazos River (shaded) proposed by Reeves (1972, fig. 2), distribution of lacustrine carbonate associated with upper and middle Blackwater Draw and upper Sulphur Draw, and existing lake basins associated with Yellowhouse Draw and upper Sulphur Draw. Boxed area is location of Figure 10A.

ward-trending stream west of Muleshoe. The ancient Portales River then flowed east. . . , bending just north of Plainview and flowing off the plains down the present White River through Blanco Canyon" (Reeves, 1972, p. 113; Fig. 10). A pre-Ogallala drainage in the Tolar-Portales area that continues eastward following the route proposed by Reeves is apparent on paleotopographic maps (Cronin, 1969, sheet 1 Gustavson et al., 1980, figs. 7, 8; Gazdar, 1981, fig. 17; Gustavson and Finley, 1985, fig. 10). There seems little doubt that portions of Running Water and Blackwater Draws follow antecedent drainages, but Gustavson and Finley (1985, p. 34) note that no investigator "provides unequivocal evidence of the actual path of the Portales River across the Southern High Plains. Thus, although the point of entry of the Portales Valley onto the Southern High Plains in eastern New Mexico is widely recognized, it is not clear whether the Portales leaves the High Plains through Blanco Canyon in the present drainage of the White River or near Yellow House

Canyon in the present drainage of the Double Mountain Fork of the Brazos River."

Evidence for an ancient drainage in lower Blackwater Draw was found in the Lubbock City Landfill (Fig. 4). In a large artificial pit (the "Wind Pit") a massive carbonate deposit (Blanco Formation) over 10 m thick is cut out along the present course of the draw (Fig. 11). Inset against the carbonate is a layer of well-sorted medium sand over 1 m thick, with planar cross-beds and some coarsening-upward sequences. The sedimentary characteristics indicate deposition by water moving in the present-day flow direction (to the southeast). The edge of the carbonate, therefore, was probably the margin of the ancestral Blackwater Draw. The bedded sand and the massive carbonate is buried by the Blackwater Draw Formation, locally up to 4 m thick. The soil in the Blackwater Draw Formation at the Wind Pit is a reddish, compound Btk horizon 2 m thick with thick, continuous clay films and a Stage II calcic horizon. These pedogenic char-

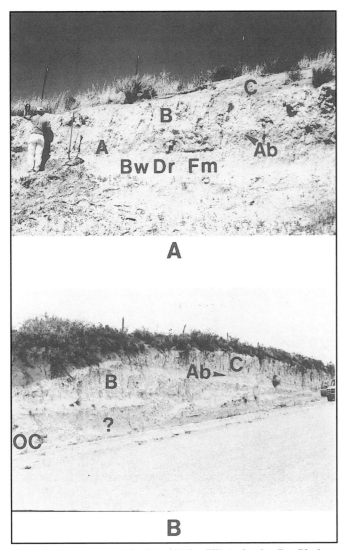

Figure 9. Photographs of the Older Valley Fill. A, Section Bw-72 along the north end of the West Bank of the Clovis site (Figs. 4, 12A), showing strata A, B, and C overlying the Blackwater Draw Formation (Bw Dr Fm). "Ab" indicates buried A horizon (dated to ca. 21,100 yr B.P.) formed in stratum A at the base of stratum B, prominent along the West and North Banks of the gravel pit. B, Broad cut at section Se-13 of the Seminole-Rose site (Figs. 4, 13B, 16) showing the Ogallala Caprock (OC), strata B (upper layer) and C, and the radiocarbon-dated buried A horizon (Ab). Strata A and the lower layer of B are difficult to differentiate in the photo and "?" indicates the zone of possible Blackwater Draw Formation.

acteristics are typical of soils in the Blackwater Draw Formation at least several hundred thousand years old (Holliday, 1989b), thus dating this reach of Blackwater Draw to at least the middle Pleistocene.

Exposures of strata A, B, and C in and around the Clovis site gravel pit on the north side of Blackwater Draw (Figs. 10A, 12A; Table 6A) provide evidence of a valley during accumulation of the upper Blackwater Draw Formation. Three layers of the Blackwater Draw Formation dip gently southward toward the draw (Fig. 12B). Soils developed in each layer show that the dip

approximates the paleotopography and, therefore, the draw was in place as the formation accumulated. Separating the individual eolian layers are deposits of carbonate, and capping the uppermost eolian layer is another carbonate (stratum B). Each carbonate layer typically is 50–100 cm thick (Fig. 9A). These carbonates are believed to be lacustrine (as opposed to pedogenic) based on their abrupt upper and lower contacts, uniform field morphologies from top to bottom, relatively low bulk density, and absence of a noncalcareous matrix similar to the overlying or underlying B horizons.

At and near the Clovis gravel pit thin layers of alluvial sand and gravel (stratum A) occur locally below stratum B and above the uppermost layer of the Blackwater Draw Formation. Stratum A becomes more extensive and thicker nearer the present-day Blackwater Draw, indicating that between lacustrine cycles or between ponded areas on the valley bottom, water flowed along the floor of the ancient basin.

In some exposures around the gravel pit the upper zone of the Blackwater Draw Formation buried by stratum B is gleyed and capped by the thick, black, A horizon of stratum A. The A horizon is cumulic, indicated by a layer of stratum A sand within it. The physical characteristics of the A horizon suggest that a well-vegetated and moist surface existed along the margins of the paleovalley prior to deposition of stratum B. The A horizon may denote a rise in the water table that produced the lake associated with stratum B. No data are available to indicate whether the gleying occurred as the A horizon formed or was the result of saturation and reduction under the lake that produced stratum B.

Stratum B crops out in extensive but discontinuous exposures on both sides of Blackwater Draw in the area of the Clovis site and for over 80 km east of the gravel pit (to the Plant X area; Figs. 4, 8). Where it occurs it typically is the surface deposit and rests unconformably on the Blackwater Draw Formation. The buried A horizon below stratum B yielded radiocarbon ages of ca. 21,100 yr B.P. (SMU-2533; Appendix 2, Table A2.1) and ca. 17,200 and 22,900 yr B.P. (C. V. Haynes, personal communication, 1993) at the Clovis site (Fig. 12B) and ca. 16,600 yr B.P. (A-6901; Appendix 2, Table A2.1) at the Jorde pit east of the Clovis site (Fig. 10A; Appendix 2, Table A2.1). Stratum A at Clovis is immediately below the radiocarbon-dated portion of the A horizon. The paleobasin containing stratum B probably was a paleovalley of the draw, connecting the Portales Valley with lower Blackwater Draw. The present middle reach of the draw, therefore, is much younger than the upper and lower reaches.

Above stratum B in the Portales Valley is a sheetlike eolian deposit of loamy to sandy sediment (stratum C) up to 130 cm thick. This deposit, Stratum II of Boldurian (1990), contains a well-expressed soil (A-Bt-Btk horizonation) that is the modern surface soil of the uplands surrounding the gravel pit (Fig. 12B).

During the late Pleistocene, therefore, upper and middle Blackwater Draw periodically was filled with an extensive hardwater marsh or lake represented by stratum B. The intercalation of stratum B with the Blackwater Draw Formation suggests cycles of transgression and regression of the water along with

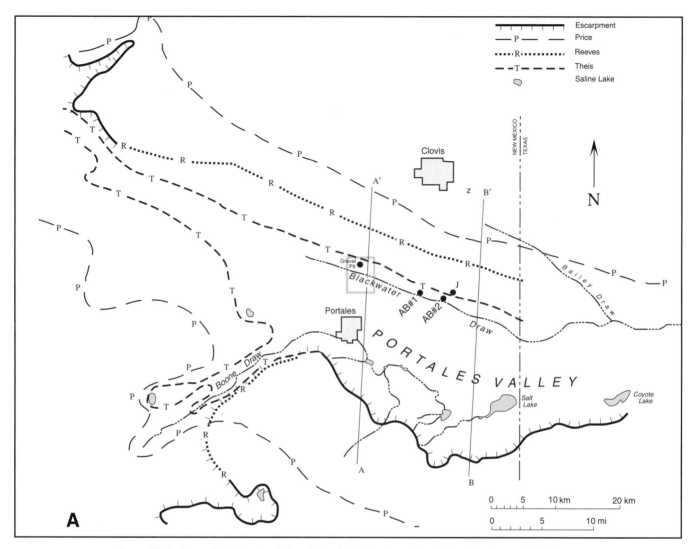

Figure 10. A, Map of the Portales Valley (Figs. 3, 8) showing: the relationship of upper Blackwater Draw and salinas to the valley; various interpretations of the limits of the valley (based on: Theis, 1932, fig. 1; Price, 1944, fig. 33; and Reeves, 1972, fig. 6); lines-of-section A–A′ and B–B′ (Fig. 10B), and locations of the Clovis gravel pit, the Anderson Basin sites (AB1 and AB2), and the Jorde pit (J). Boxed area is location of Figure 12A. B (on facing page) , Topographic profiles A–A′ and B–B′ (Fig. 10A) through the greater Portales Valley showing the relationship of Blackwater Draw to the Inner Portales Valley.

cycles of eolian sedimentation (deposition of the Blackwater Draw Formation) when the water was down. The degree of pedogenic development in the buried soils between carbonate layers (Appendix 1, Table A1.8) further suggests that the sequence spans tens of thousands, if not hundreds of thousands of years (Holliday, 1989b, 1990b).

One characteristic of the Portales Valley not noted by other investigators is that it appears to be the headwaters of two draws. As noted above, Blackwater Draw is along the northern margin of the Portales Valley and extends eastward from the Clovis-Portales area to a point about 65 km (40 mi) east of the state line and then turns south (Fig. 8). The Inner Portales Valley begins west and southwest of the city of Portales and drains southeast-

erly through Portales and into the Salt (or Arch) Lake basin, which in turn drains into the Coyote Lake basin (Fig. 10A). Filling of the Coyote basin would result in overflow into the North Fork of Yellowhouse Draw (Fig. 8). The Portales Valley, therefore, is the head of Blackwater Draw and possibly North Yellowhouse Draw.

South of the Portales Valley is the Simanola Valley, another broad, shallow depression on the High Plains surface that, like the Portales Valley, begins as a reentrant on the west side of the Llano Estacado (Fig. 3). Also like the Portales and Brazos River valleys, the Simanola Valley may mark the location of a paleovalley of the Colorado River that drained to the east or southeast prior to integration of the Pecos system. Below the present

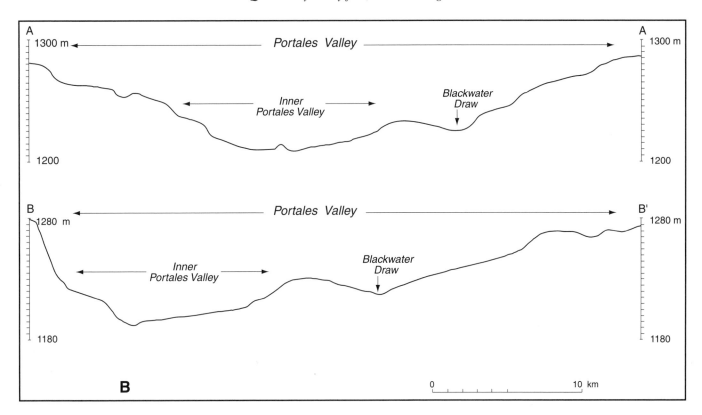

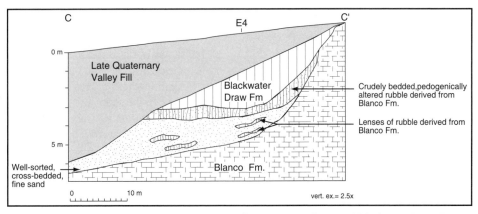

Figure 11. Generalized geologic cross section C–C′ through the Blanco and Blackwater Draw Formations at the Lubbock Landfill in lower Blackwater Draw (Figs. 4, 27A) with the location of section E4.

Simanola Valley, on the pre-Ogallala erosion surface, is "an ancient valley cut about 30m into the Triassic bedrock, and about 25–45m wide . . ." (Gazdar, 1981, p. 45). The location and orientation of this buried valley is not indicated by Gazdar (1981, fig. 11), however, and paleotopographic maps of the pre-Ogallala surface in Texas (e.g., Cronin, 1969, sheet 1) show no indication of a paleovalley. Reeves (1972, p. 112) refers to the "Simanola Valley–Ranger Lake drainage" as post-Pliocene, but does not elaborate. The absence of evidence for the Simanola Valley on the pre-Ogallala surface indicates, however, that this valley may be younger than the Portales Valley. Gustavson and Finley (1985, fig. 13) suggest that Sulphur Draw or Sulphur Springs Draw or

all or parts of both may follow the Simanola Valley, but there are no data to provide any indication of the route of the ancient Simanola Valley (ancestral Colorado River).

Deposits of stratum B are extensive in upper Sulphur Draw from the Bledsoe site to Milnesand (Figs. 4, 8). The carbonates appear to fill a broad, shallow paleobasin in the headwaters of the draw. A relationship between this paleobasin and the Simanola Valley is not apparent. Sulphur Draw also is associated with a number of extant lake basins. In the headwaters area of the draw there are 17 playas, aligned roughly parallel to and mostly south of the draw (Fig. 8). These playas appear to be deflated into stratum B.

Another large reentrant, the Winkler Valley (D. J. Meltzer, personal communication, 1990), occurs on the southwest side of the Llano Estacado (Fig. 3). There is no evidence that this valley is related to the draw systems of the Colorado. The Winkler Valley differs from the Portales and Simanola Valleys in trending northeast-southwest, and has no apparent physiographic relationship to the present draw systems. Data are not available to indicate whether there were Tertiary drainages through this area.

Older Valley Fill was identified in Yellowhouse, Seminole, Sulphur Springs, Mustang, and Monahans Draws, in addition to Blackwater and Sulphur Draws. At the Lupton site on lower Yellowhouse Draw (Figs. 4, 13A), stratum B, almost 2 m thick, was

isolated by downcutting of the drainage. Organic-rich sediment within the carbonate produced a radiocarbon age of ca. 17,700 yr B.P. (SMU-2235; Appendix 2, Table A2.1). At the Seminole-Rose site on lower Seminole Draw (Figs. 4, 9B, 13B), stratum B, almost 4 m thick, interfingers with stratum A, documenting fluctuations in the extent of the ponds or marshes that produced the carbonate. Stratum B also was isolated along the valley wall by downcutting. An organic-rich layer within the lake deposit is dated to ca. 16,300 yr B.P. (SMU-2342; Appendix 2, Table A2.1). Augering the uplands adjacent to the Midland site on Monahans Draw revealed possible equivalents to strata B and C, and Meltzer (1991) reports upland sheet sands (stratum C?) adjacent to lower Mustang Draw.

The Older Valley Fill also is preserved near the mouth of Sulphur Springs Draw (Palmer-Wheeler area, Fig. 4). Frederick (1994) reports late Pleistocene deposits preserved in a terrace (T2) of the valley: gravel and sand (stratum A?), marl (stratum B?), and eolian sand heavily modified by pedogenesis (stratum C?). Eolian sheet sand (stratum C?) also is extensive throughout the uplands around the mouth of the draw (Frederick, 1993a).

The dated sections of Older Valley Fill show that the final downcutting in Blackwater, Yellowhouse, and Seminole Draws probably did not occur until at least about 15,000 yr B.P. given that the radiocarbon ages of stratum B predate the final stage of lake sedimentation. The incision ended between 12,000 and 11,000 yr B.P., based on dating of the Younger Valley Fill.

Geomorphic characteristics and hydrogeology

The draws of the Brazos and Colorado River systems have varying genetic histories, but they share several physical characteristics. All draws head in New Mexico and generally follow the regional slope to the southeast. They are incised into the Blackwater Draw Formation and locally into older deposits (Table 4). Most draws are broad and relatively shallow (Table 5), with gen-

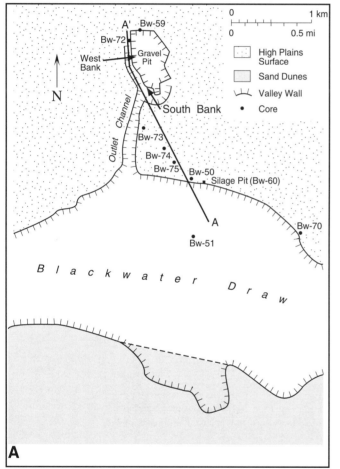

A

Figure 12. A, Upper Blackwater Draw in the area of the Clovis site (Figs. 4, 10A) showing: the location of the gravel pit at the site (including the South Bank section at the north end of the "outlet channel" and the West Bank); location of cores and sections studied in the area; and line-of-section A–A' (Fig. 12B). Key is for all site maps. B, Geologic cross section A–A' through the Clovis site (Fig. 12A) and into Blackwater Draw illustrating stratigraphy of the Older Valley Fill and the generalized upper Cenozoic record.

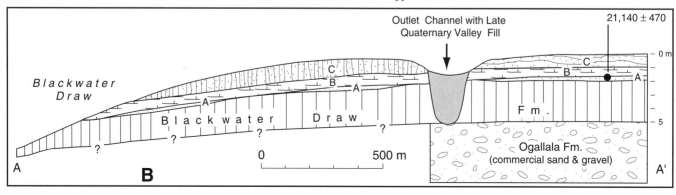

B

**TABLE 6A. CORRELATION OF SELECTED STRATIGRAPHIC TERMINOLOGY
FOR THE CLOVIS SITE, BLACKWATER DRAW***

Stock and Bode 1936	Sellards and Evans 1960	Haynes 1975†	Haynes 1995†	Holliday This Volume
Eolian sand	`Tan Eolian Sand	Units G1, G2	Units G2, G3	Stratum 5s
	Jointed Sand	Unit F	Unit G1	Stratum 4s
			Unit F	Stratum 3s
Blue sand	Carbonaceous Silt	Unit E	Unit E	Stratum 2s
	n.r.	Unit D	Units D0, D2z	Stratum 2s
	Diatomaceous Earth	Unit D	Units D1, D2	Stratum 2d
Caliche?	Brown Sand Wedge	Unit C	Units D2x, D2y	Strata 1/2
Speckled sand?	Gray Sand	Unit B1	Unit C	Stratum 1s2
Speckled sand	Gray Sand?	Unit B1	Unit B1-B3	Stratum 1s1
n.r.	n.r.	n.r.	Unit A9	Stratum C§ Stratum B**
n.r.	n.r.	n.r.	n.r.	Stratum A
n.r.	n.r.	Units A5-A13, B2	Units A3-A8	Blackwater Draw Fomation
Yellow sand	Bedrock Gravel	Units A1-A4	Units A1, A2	Ogallala Formation (Commercial sand and gravel)

*Figures 4, 10A, 12A. Modified from Haynes, 1975, Table 4-1.
†Stratigraphic nomenclature primarily applies to "South Bank" and "West Bank."
§Stratum II of Boldurian, 1990.
**Stratum I of Boldurian, 1990.
n.r. = strata not recognized.

tly sloping valley walls and flat floors underlain by late Quaternary sediments (Fig. 14). The associated drainage basins are very narrow; generally equivalent to the widths of the draws. Beyond the valley walls the High Plains surface is very level and provides little runoff into the draws.

Apart from general similarities, however, there are some significant differences in geomorphic characteristics among and even along the draws. The mean slopes (channel gradients) of most draws are from 1.6–1.9 m/km, although the slopes for Blackwater and lower Mustang are considerably less (Table 5). These differences in valley slope are because Blackwater and lower Mustang do not follow the regional slope: upper Blackwater flows essentially to the east, and lower Blackwater and lower Mustang have a strong southward (rather than southeastward) component to their flow directions. The absolute dimensions of each valley vary considerably (Table 5). Running Water and lower Blackwater Draws are the largest by all standards: >200 km long and typically 500–1,000 m wide and 10–20 m deep. The tribu-

tary draws of the Mustang system (above the confluence with Mustang proper) are shorter (70–160 km), and all draws of the Mustang system are more narrow and shallow (typically 100–400 m wide and ≤15 m deep). Yellowhouse Draw is intermediate in size: 175 km long; widths generally 200–600 m; depths are typically ≤10 m.

More significant than absolute dimensions, however, is the range in variation along each draw. Running Water and Blackwater Draws, especially the latter, vary considerably in width and depth (Table 5), whereas the draws of the Colorado system are much more uniform in these dimensions as well as being smaller (Table 5). Several factors probably are responsible for these differences along and between the draws, but one that may be particularly significant is the variable thickness of the Blackwater Draw Formation. In the Running Water–Blackwater area the formation typically is 5–10 m thick, whereas in the area of the Mustang system the deposit is usually <5 m thick and most commonly 1–2 m. In the thicker Blackwater Draw Formation the valleys can

**TABLE 6B. CORRELATION OF STRATIGRAPHIC TERMINOLOGY
FOR THE AREA OF ANDERSON BASIN, BLACKWATER DRAW***

Anderson Basin 1		Anderson Basin 2		
Haynes 1975	Holliday This Volume	Stock and Bode 1936	Haynes 1975	Holliday This Volume
Unit G2			Unit G	Stratum 5s
Unit G1	n.r.	Eolian sand		
Unit F			Unit F	Stratum 4s
Unit E	Stratum 2s	Brown sand	Unit E	Stratum 2s
Unit D	Strata 2s and 2d	Blue sand	Unit D	Strata 2d and 2s
Unit C	Stratum 1	Caliche	Unit B1	Stratum 1
Units B1 and B2	n.r.	Yellow sand	Units A6-A13	n.r.

*Figures 4, 10A, and 23.
n.r. = strata not recognized.
See notes following Table 6D for lithologic and pedologic descriptions.

**TABLE 6C. ARCHAEOLOGICAL TRADITIONS AND ASSOCIATED (SELECTED) TERRESTRIAL
VERTEBRATE FAUNAS REPORTED FROM UPPER BLACKWATER DRAW**

Stratum	Archaeological Tradition	Vertebrate Fauna	Sites*
5	Late Prehistoric	None reported	Clovis†
4	Archaic	*Bison bison*	Clovis†
2s	Late Paleoindian	*Bison antiquus*	Clovis† and Anderson Basin§
2d	Folsom	*Bison antiquus*	Clovis** and Anderson Basin**
1	Clovis	*Mammuthus columbi, Camelops* sp. *Equus mexicanus* and *E. Francisi, Bison antiquus, Smilodon californicus, Canis dirus, Capromeryx* sp.	Clovis** and Anderson Basin‡

*Figures 4, 10A, and 23.
†Hester, 1972.
§Hester, 1975.
**Stock and Bode, 1936; Hester, 1972; Lundelius, 1972; Slaughter, 1975; Johnson, 1986.
‡Fauna only; Howard, 1935a; Stock and Bode, 1936.

widen relatively easily because the formation is easily eroded by flowing water. In the south, however, the valleys are incised into much more resistant layers below the Blackwater Draw Formation; layers that would significantly constrain widening. Data from Yellowhouse Draw support this hypothesis. The draw tends to be much wider above the Narrows (Fig. 4), corresponding to an area of thicker Blackwater Draw Formation, whereas below the Narrows the formation is thinner and overlies more resistant calcretes and lacustrine carbonates.

The relatively large size of the Running Water and Blackwater systems also may be related to the presence of the ancestral Brazos River or other large, antecedent valleys. Such ancient rivers would leave large, topographic and sedimentologic "pathways" easily inherited by subsequent drainage systems (i.e., the draws).

Blackwater Draw has the most striking variations in morphology of any draw and accounting for these differences is difficult. The lower reach of the draw (beginning at about the Gibson

TABLE 6D. LITHOLOGIC AND PEDOLOGIC CHARACTERISTICS OF BLACKWATER DRAW

Stratum/ Soil	Range in Thickness (cm)	Description
Unnamed soil		A (ochric or mollic) - Bw; or A (ochric or mollic) - Bw/Bk or Btj/Bk (Stage I calcic).
5	69 - 240	**5s:** SL (Bw-17); S in dunes.
	70	**5g:** S and bedded gravel.
	30 - 80	**5m:** SL, SiL
Lubbock Lake soil		**Valley-axis:** A (mollic) - Bt/Btk (argillic; Stage I-II calcic); A is locally cumulic due to slow deposition of stratum 5m. **Valley-margin:** A (ochric) - Bt/Btk (weak argillic; Stage I-II calcic).
4	40-265	**4s:** SL, LS; clayey locally; sandy valley-margin facies, especially in and near dunes; divided into 4s1 and 4s2 at Lubbock Landfill.
Yellowhouse soil		**3c:** A (ochric, locally cumulic) - C
3	20-200	**3c:** SiC, CL, SC, SCL, SL; 9-38% carb; local lenses of low-carbonate sand. **3s** S, SL; divided into 3s1 and 3s2 at Lubbock Landfill; OM-rich lenses locally common.
2	100-200	**2s** (Clovis, Gibson): fS (ca. 100 cm). **2m** (Tolk): interbedded mud and fS (22 cm). **2m** (Davis): LSi (30 cm). **2m** (Progress): LSi (140 cm). **2m** (Anderson Basin #1): SiL (17 cm). **2m** (Clovis): OM-rich SL-L with some diatoms; interfaces with 2s at basin margin (20 cm in basin to 60+ cm at margin). **2d** (Anderson Basin 1 and 2): diatomite (10-20 cm). **2d** (Clovis, Gibson, L. Landfill): interbedded diatomite and mud; local lenses of fS.
1	0->250	Massive, fine, quartz sand (with local cross-beds or contorted bedding) over m-c gr (locally bedded); interbedded C (1-5 mm) locally common.

Notes for lithologic and pedologic descriptions used in this table and Tables 7C, 9C, 10, 11C, 12, 13C, and 13D:
Texture = carbonate-free basis.
Textural abbreviations: v = very; f = fine; m = medium; S = sand; C = clay; L = loam; SC = sandy clay; SCL = sandy clay loam; SL = sandy loam; LS = loamy sand; CL = clay loam; SiL = silty loam; SiCL = silty clay loam; gr = gravel; f = fine; m = medium; c = coarse.
OM-rich = organic-matter rich.
Carb for 3c = carbonate content of unweathered marl.
Stratum 2 lithology is described for individual sites because of its relative rarity and variability.
For pedologic characteristics, basic soil horizonation is provided with diagnostic horizons if appropriate.
...cm = local thicknesses.

site, Fig. 4) is very large; up to 1,768 m wide and 88 m deep. Updraw, between the Gibson site and the state line, however, the draw is much less pronounced and locally is difficult to follow. At the Tolk site the valley is only 274 m wide and 6 m deep. Above the state line the draw is again larger and becomes both deeper and wider as the Portales Valley becomes more pronounced. The narrow, shallow, middle reach is inset into extensive deposits of late Pleistocene lacustrine carbonate (stratum B) probably deposited in a paleovalley of the draw (discussed above). The middle reach, therefore, probably is much younger than the upper and lower reaches.

Upper Running Water Draw also is significantly wider at its upstream end (>2,000 m wide and 27 m deep at the Houck site and <1,000 m wide and <20 m deep at most other sites). Upper Running Water and Upper Frio Draws are straight and parallel, suggestive of structural control. Structural features such as joints and fractures could facilitate erosion or produce localized subsidence and account for the unusual depth of upper Running Water.

The drainage patterns of the draw systems have few characteristics in common with patterns typical of perennial, integrated streams and the patterns also vary among draws. Running Water, Sulphur, and Sulphur Springs Draws flow southeasterly in linear to rectilinear patterns and have few tributaries. Blackwater Draw flows easterly then southeasterly in both linear and curvilinear patterns and also has few tributaries. Yellowhouse Draw flows roughly eastward, but is composed of several tributaries linked by lake basins and has a curvilinear drainage pattern. The draws of the Mustang system differ considerably from the draws to the north in being composed of several tributary valleys (McKenzie, Seminole, and Wardswell Draws) draining into the Monument-Mustang system and forming a dendritic drainage pattern (Fig. 3). These variations in drainage patterns are probably because of the relatively short time available for drainage development (most of the landscape dates to the late Pleistocene at the oldest) and controls exerted on development of the draws by subsidence and antecedent topography, discussed above.

At the scale of individual valleys and sites, a noteworthy characteristic of some draws is entrenched meanders, which are particularly common along lower reaches of Running Water, Blackwater, and Yellowhouse Draws, and throughout Sulphur, Seminole, and Mustang Draws (Fig. 15). A few reaches have cut-off or near-cutoff meanders, although other reaches are essentially straight. These entrenched meanders do not compare in sinuosity to such classic examples as those of the Colorado Plateau, but nevertheless are obvious on topographic maps and aerial photographs (Figs. 14C, 16). Entrenched meanders commonly are associated with plateaus or horizontal bedrock (Blache, 1940, in Harden, 1990; Gardner, 1975) and low channel gradients (Harden, 1990), and appear to be related to either vertical uplift or base-level lowering or both (Gardner, 1975). The High Plains meets these topographic and geologic conditions. The region is a plateau composed of horizontal layers of rock and has low topographic gradients (1.0–2.0 m/km). The High Plains probably was uplifted 1,000–1,500 m over the past 10 million

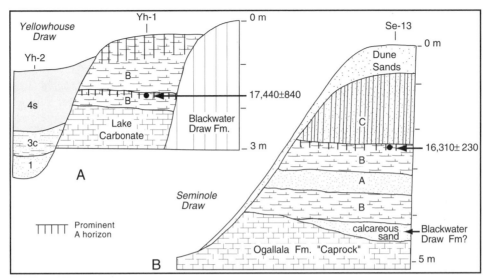

Figure 13. Schematic geologic cross sections of: A, the Lupton site (from core Yh-1 to Yh-2; Fig. 4); and B, Pleistocene lake sediments exposed at the Seminole-Rose site (see also Fig. 9B; Fig. 4). No horizontal scale. See Figure 6 for key to symbols.

years (Gable and Hatton, 1983). Base level also was lowered throughout the Pleistocene as the eastern escarpment of the High Plains retreated westward and the Brazos and Colorado River systems incised into the Rolling Plains. There are no data, however, to indicate what influence each process had on development of the incised meanders of the High Plains.

Today there is no flowing water in the draws, and standing water is rare although ponding occurs locally after heavy rains. Historically, however, some draws carried significant amounts of water, supported by springs and seeps until the advent of ground water withdrawal in the first half of this century (Brune, 1981; GSA Data Repository 9541, Appendix D). The springs and seeps were fed by the Ogallala aquifer or possibly by local, perched water tables, but no studies are available that describe the hydrogeologic relationship of the aquifers to the surface drainage. As described below, some reaches of the draws had numerous springs and carried water perennially, but other reaches remained dry year-round. The hydrogeologic conditions that determined the distribution of springs is unknown. One controlling factor may be the Ogallala Caprock caliche. The Caprock is known to vary in thickness and permeability, and locally is discontinuous (Gustavson and Holliday, 1988). Data on bedrock geology are available from only three localities of documented spring activity in draws: Clovis (Haynes and Agogino, 1966; Haynes, 1975, 1995) Lubbock Lake (Johnson, 1974; Stafford, 1981; Holliday, 1985c) and Mustang Springs (Meltzer, 1991). At all three sites the Ogallala Caprock is absent. Other important factors may be the depth of valley incision relative to the elevation of the water table. Comparisons of these variables are not available, however.

Based on historic documentation, there were significant differences in spring activity and stream flow along the draws (Fig. 17). Running Water Draw was obviously named for its hydro-

logic characteristics and according to Brune (1981) springs fed ponds and creeks along the middle reach of the draw at and above Plainview in the early twentieth century (Fig. 17; GSA Data Repository 9541, Appendix D). Middle Blackwater Draw had substantial spring-fed ponds and stream flow such that "in early settlement days" people were able to boat from town to town (Brune, 1981, p. 58, 284; Figs. 17, 18). Springs and seeps also were common along lower Blackwater Draw, beginning below the Ben site (Fig. 4; Brune, 1981, p. 298; Bolton, 1990, p.

Figure 14. Photographs illustrating both the general characteristics and the morphological variability among the draws. A, Upper Running Water Draw (view north) at the Ned Houck site (Fig. 4; with truck at Rw-4). This reach of the draw has more relief than any other reach of any draw in the Brazos or Colorado systems. B, Upper Blackwater Draw (view south, taken from the north valley wall) at the Davis site (Fig. 4; with truck at Bw-44). The width and relief are typical of the draw below the Anderson Basin area. Just below the skyline, above and right of the coring mast, is an outcrop of Pleistocene lake carbonate. C, Lower Yellowhouse Draw at Lubbock Lake (Figs. 4, 28A), looking west. The trees at left center outline the old reservoir where most of the archaeological excavations focused. Note the entrenched meander at right and the white outcrops of the Blanco Formation along the valley walls (photo provided by Lubbock Lake Landmark, The Museum of Texas Tech University). D, The Edwards Road site on upper McKenzie Draw (Fig. 4), looking north-northwest, with the truck at Mk-12. The draw is shallow and relatively narrow here. The white outcrops in the foreground and on the opposite side of the draw are the Ogallala Caprock Caliche, which is very near the surface in this area of the High Plains. E, Mustang Springs on lower Mustang Draw (Figs. 4, 30A) looking west (taken from low on the east valley wall). The draw has a flat floor and steep valley walls in this reach. The area of dark vegetation at left center is the depression caused by erosion from historic spring discharge (Fig. 30). Mounds of earth at right mark backhoe trenches that provided most of the data for the section in Figure 30B.

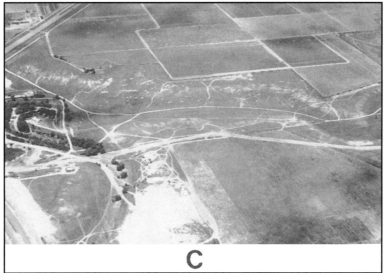

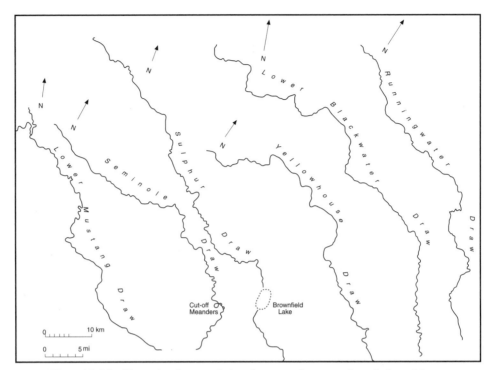

Figure 15. Map illustrating the meandering character of most reaches of selected draws.

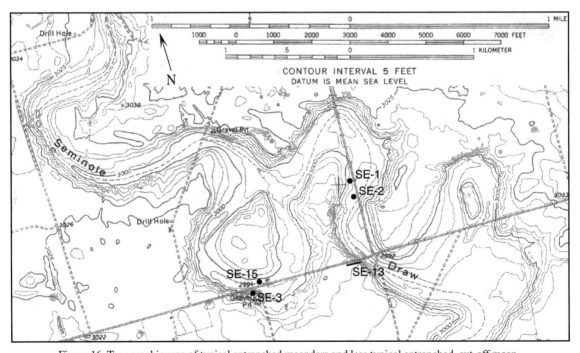

Figure 16. Topographic map of typical entrenched meanders and less typical entrenched, cut-off meanders on lower Seminole Draw above and below the Seminole-Rose site (Fig. 4; from the McKenzie Lake SE Quadrangle, 1:24,000). Cores and sections for the Seminole-Rose site (Se-1, Se-2, Se-13) and Cutoff Meander site (Se-3, Se-15) also are located.

273). The Spanish also report springs, used as landmarks, along middle Blackwater Draw and in the Portales Valley (Bolton, 1990, p. 273). Standing and flowing water is reported by Brune (1981) and Holden (1959, 1974) on Yellowhouse Draw, mostly at and below Lubbock Lake (considered the headwaters of the Brazos River; Holden, 1974), scattered along Sulphur Draw and upper and middle Sulphur Springs Draw, on lower Mustang Draw, mostly at and below Mustang Springs, and in a few localities along Wardswell, McKenzie, Seminole, and Monahans Draws (Fig. 17; GSA Data Repository 9541, Appendix D). The largely discontinuous nature of the flow along the draws indicates that the water seeped back into the ground after flowing away from the scattered spring-fed ponds and lakes. White et al. (1946, p. 387) noted that peak flood flow in May 1937 along Running Water Draw at Plainview was 1,200 cubic feet per second while the maximum flow 15 miles downstream was 80 cubic feet per second.

Stafford (1981) proposed that spring-fed ponds and marshes may be associated with deeply entrenched, highly sinuous meanders of draws, based mostly on geoarchaeological studies in lower Yellowhouse Draw at and below Lubbock Lake. The present study provides no evidence of a relationship between the location of springs and entrenched meanders. The highest concentrations of springs (middle Running Water Draw, middle Blackwater Draw) are not along highly sinuous reaches. Indeed, the two best-known spring sites other than Lubbock Lake, Clovis (Haynes and Agogino, 1966), and Mustang Springs (Meltzer, 1991; Figs. 2, 4), are not associated with entrenched meanders. The reaches where entrenched meanders are most common (lower Running Water, lower Blackwater, lower Yellowhouse, and throughout Sulphur, Seminole, and Mustang Draws) have relatively few springs or none at all (compare Figs. 15 and 17).

Active springs no longer occur in the draws, except briefly following unusually wet years (Cronin, 1964; Fallin et al., 1987; Ashworth, 1991; Ashworth et al., 1991). Marshes sustained by high ground water can be found in a few areas, however. The most extensive areas are in lower Sulphur Draw, below Brownfield Lake (Fig. 4), and lower Sulphur Springs Draw, where the modern channel incised the valley fill and bedrock (Quigg et al., 1994). Marshy conditions hampered coring at New Moore West and New Moore South on Sulphur Draw (Fig. 4) and prevented sampling farther down valley. The ground water in this reach of Sulphur Draw is "moderately saline" (Brune, 1981, p. 427) and not used for irrigation. The water table remains high, therefore, and maintains the extant marshes (Brune, 1981, p. 427). The waters in lower Sulphur Springs Draw also are saline and the extant lakes and ponds on the floor of the draw near its mouth are described as "salt lakes" (Quigg et al., 1994, p. 5).

STRATIGRAPHY OF THE VALLEY FILLS

This section is a data presentation and discussion of the lithostratigraphy, pedostratigraphy, and geochronology of the Younger Valley Fill. These deposits are largely continuous latest Pleistocene and Holocene sediments that filled the draws following the final phase of downcutting.

The Younger Valley Fill includes the strata and data reported from scattered sites along the draws prior to 1988 (see earlier section, "Introduction"), as well as information gathered from 1988 to 1992 resulting from field work in addition to the author's study sponsored by the National Science Foundation (Brown et al., 1993; Johnson, 1994; Quigg et al., 1993, 1994). The classic stratigraphic sequence known prior to 1988 includes, from bottom to top: latest Pleistocene alluvium; latest Pleistocene and earliest Holocene paludal mud (with diatomite and diatomaceous mud); early Holocene paludal carbonate; early to middle Holocene eolian sediment with a relatively well-expressed soil formed in the late Holocene; and local occurrence of latest Holocene eolian, slope wash, and paludal sediments with weakly to moderately expressed soils (e.g., Sellards and Evans, 1960; Haynes, 1975; Stafford, 1981; Holliday, 1989a; Meltzer, 1991). The 1988–1992 study indicates that, in general, this stratigraphic sequence occurs throughout most of the draws. The total thickness of the fill typically is between 2 and 4 m, although thinner and thicker sections were found (Appendix 1). Ubiquitous units include a basal sand and gravel, a carbonate layer, and an overlying sandy to loamy fill with a moderately to well-expressed soil profile. There are two exceptions, however, to the classical view of the stratigraphy based on the recent field studies: above the sand and gravel the muds and diatomites occur only very locally, and the geochronology of these strata varies significantly from locale to locale.

The lithology and stratigraphic relationships of the valley fill are sufficiently uniform along and among the draws to allow definition of an informal group of lithostratigraphic units, based on those established at the Lubbock Lake site (Johnson, 1974; Stafford, 1981; Holliday, 1985c). Five lithostratigraphic units are defined: strata 1–5, from bottom to top (Figs. 6, 19, 20, 21, 22). Strata 2–5 include several significant facies indicated by suffixes: c = calcareous (primary carbonate); d = diatomaceous (including diatomite); m = muddy (carbonaceous mud); s = sandy, loamy, or sandy and gravelly; si = silty; g = gravelly. Lithofacies nomenclature includes terms common in either geology or pedology, but not both; in particular "mud" and "loam" are used in accordance with conventions of the respective disciplines and are not interchangeable. These terms are used for convenience and brevity. A mud is a deposit containing mostly silt and clay (after Bates and Jackson, 1980). Muds in the valley fill typically are stained dark gray or black by organic matter. A loam is a deposit with more or less equal amounts of sand, silt, and clay (Soil Survey Division Staff, 1993). There are two named pedostrata: the Yellowhouse soil and the Lubbock Lake soil (Figs. 20F, 22). Chronostratigraphic conventions are as follows: the Pleistocene-Holocene boundary is 10,000 yr B.P. (after Hageman, 1972), early Holocene 10,000–7,500 yr B.P., middle Holocene is 7,500–4,500 yr B.P., and late Holocene is 4,500–0 yr B.P. (Fig. 19).

The sandy, generally valley-margin facies of strata 2, 3, and

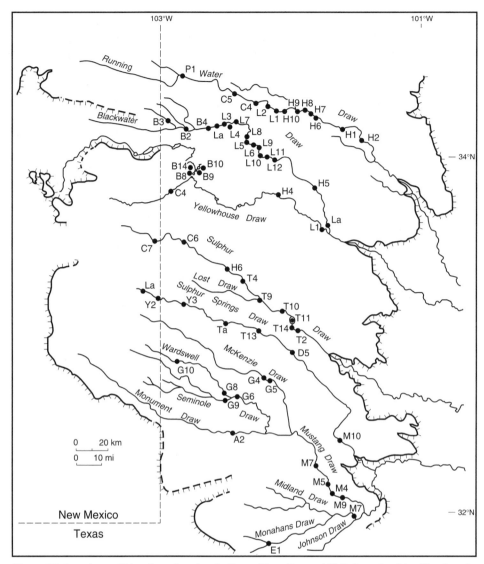

Figure 17. Locations of historic spring sites in Texas (from Brune, 1981) (keyed to identifications in
GSA Data Repository 9541, Appendix D).

4 are very similar to one another, but usually could be differentiated. Designation of a sandy facies as stratum 4s was based on
clear evidence of stratigraphic superposition above stratum 3
(any facies) and lack of evidence for a facies relationship with
3c. Stratum 4s probably included 3s at some localities where the
two units were welded by formation of the Lubbock Lake soil.
Similarly, designation of a sandy facies as stratum 3s was based
on clear evidence of burial below stratum 4 (any facies) and evidence for a facies relationship with 3c. Stratum 2s, which is rare,
was identified where a sandy layer exhibited a clear facies relationship with stratum 2m or 2d. Some sandy units identified as 3s
probably include 2s. Such terminological problems are not considered significant, however, because the environmental discussions and reconstructions are based on comparison of lithofacies.
    Very localized facies of strata 1–5 or other localized deposits

also were encountered along the draws. These deposits are of
such limited extent that they were given no stratigraphic identification, although they are described where necessary. Also common along the draws are very recent accumulations of sandy
sediment, probably resulting from erosion of the landscape following the introduction of agriculture. These deposits are considered to be modern based on the complete absence of evidence
of weathering.
    The strata are more or less continuous along all draws, with
a few exceptions. Yellowhouse Draw connects a series of large
saline playas and several small playa basins (see previous section, "Geologic and Geomorphic Background"). As a result, late
Quaternary sediments occur in segments between lake basins.
Upper Sulphur Draw in New Mexico consists of a series of
broad, interconnected lake basins partially obscured by sand

Figure 18. Blackwater Draw at the Gibson site (Fig. 4) in 1951. The photos were taken from the crest of the dunes bordering the south side of the Marks Beach blowout (Fig. 29) looking southeast (left) and south-southeast (right), showing the spring-fed pond in the floor of the draw (right; photos courtesy F. E. Green). Here the draw slopes east to west (left to right), the only such reach identified for any draw in the region. Settings similar to this one (ponds along the valley axis, dry sands at the valley margins, and soil formation between the two areas) probably were common in the early Holocene.

dunes and lunettes (see previous section, "Geologic and Geomorphic Background"). In lower Sulphur, below Brownfield Lake, coring was seriously hampered by a high water table that kept the valley fill saturated and locally intersected the floor of the draw to produce saline ponds. In lower Sulphur Springs Draw the younger valley fill is exposed in a terrace created by recent incision of the drainage (Frederick, 1994).

No single type section is proposed for the younger valley fill because there are few exposures (most of the research was done from cores) and because no formal stratigraphic scheme is proposed. Reference sections for future studies, however, are found at the Lubbock Lake site, where the lithostrata and pedostrata were first described, the Plainview site (Pit 3), the Clovis gravel pit, the Brownfield site, and the Glendenning pit (Fig. 4). Samples from extensive trenching at Lubbock Lake are housed at the Museum of Texas Tech University. Samples collected by the author at other sites, including the 1988–1992 coring, are stored in the Geography Department at the University of Wisconsin.

Stafford (1981) proposed formal lithostratigraphic nomenclature for the valley fill. The Yellowhouse Formation includes fluvial and lacustrine sediments of strata 1, 2, and 3, and the Lubbock Formation is composed of fluvial, eolian, colluvial, and marsh sediments of strata 4 and 5. The formations are separated by an unconformity between strata 3 and 4. Dividing the valley fill into two formations is not considered a usable concept based on data collected subsequent to Stafford's research (Holliday, 1985a, b, c; this paper). This is because of the significant amount of eolian sediment now documented for strata 2 and 3 and because of the presence of lacustrine sediment in stratum 4. The unconformity between strata 3 and 4 also is localized.

Summaries of the general range in lithologic and pedologic characteristics of each strata within each draw are presented in tabular form (Tables 6–13). Correlations of the newly proposed lithostratigraphic nomenclature with the schemes from selected previous investigations also are offered along with summaries of archaeological associations and selected terrestrial vertebrate faunas (Tables 6–9, 11, 13). Descriptions of individual sites, cores, and exposures, and all laboratory data and radiocarbon ages are presented in Appendixes 1 and 2, and in Appendixes A and B in the GSA Data Repository 9541.

Stratum 1

Stratum 1 typically is gravel, composed of silicified calcrete, and bedded fine sand composed of quartz. The deposit usually is 1–2 m thick, but the full range of lithologies and thicknesses is unknown because coring and augering through this deposit was very difficult. Core barrels could not penetrate the gravel and the drill hole usually collapsed once augering began. Stratum 1 was found at almost all sites. The deposit usually is deeply buried, but in upper Blackwater Draw, above Anderson Basin #2, stratum 1 is exposed on the floor of the draw due to deflation of overlying deposits. In a few exposures stratum 1 was observed on terraces

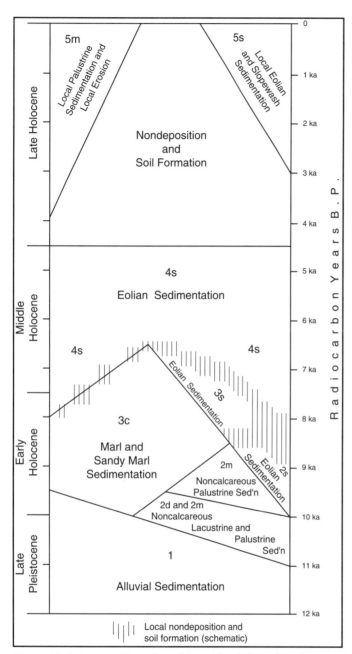

Figure 19. Schematic illustration of depositional chronology (vertical scale) and relative dominance of depositional environments (horizontal scale). Strata 3m, 4m, and 5g are localized and not included.

above the bedrock floor of the draws, but usually was buried by younger valley fill.

There are a few lithologic variations in stratum 1. Locally the deposit ranges in texture from loamy sand to sandy clay, and in some sections clay lenses are common. At Anderson Basin #2, the sands in the upper 20–30 cm of stratum 1 are heavily stained by organic matter (the "dark gray sand" of Hester, 1975, fig. 2–25; Fig. 23). This zone is an A horizon developed in stratum 1 and represents the establishment of a stable (i.e., nonaggrading) and perhaps marshy landscape denoting the onset of conditions

that resulted in deposition of stratum 2. Frederick (1994) notes the occurrence of secondary gypsum in stratum 1 in lower Sulphur Springs Draw.

Valley-margin facies of stratum 1 were found at two sites. The pit at Glendenning (lower Mustang Draw, Fig. 4), exposed poorly sorted, unbedded layers of coarse, angular calcrete fragments mixed with rounded and better sorted gravel. The valley-margin facies of stratum 1 at Midland (Monahans Draw; Figs. 4, 24B) includes calcareous silty and clayey loams, interbedded with thin lenses of low-carbonate sands and gravels. Some of the more calcareous zones appear to be composed of nodules and fragments derived from a lacustrine calcrete exposed in the valley walls (Fig. 24B).

Sandy, valley-margin facies of stratum 1 are identified in the ancient basin of the Clovis site and are interpreted as spring-laid deposits (e.g., Haynes, 1975; Haynes and Agogino, 1966; Haynes et al., 1992; Fig. 25). The deposits are of considerable archaeological significance because they are associated with Mammoth and *Bison antiquus* remains and artifacts of the Clovis and Folsom cultures. The stratigraphic relationships of these sands, particularly the "brown sand wedge" of Sellards (1952), are unclear and remain somewhat controversial, however, owing to continued quarrying during various phases of investigation (e.g., Green, 1992; Haynes et al., 1992). At least some of these sands are facies of both strata 1 and 2 (hence the inclusion of both Clovis and Folsom artifacts; Haynes, 1995; Fig. 25; Table 6A). Valley-

Figure 20. Photographs of sections that illustrate some of the more common sedimentary facies and soils found in the valley fill. A, A section at the Plainview site (Fig. 4), looking northwest at the south end of the west wall of Pit 3 around Section 7 (arrow; Fig. 26). The Ogallala Caprock (OC) comprises the lower half of the section. The late Pleistocene muds (1) inset into an old channel cut in the bedrock are visible as are the eolian facies of strata 3 and 4. B, The lower section of Rw-19 at the Edmonson site (fig. 4; bottom of core to the left), showing bedded sands (stratum 1), interbedded muds and sands (stratum 2m), marsh carbonate (stratum 3c), and black muddy A horizon formed within 3c (3Ab2). C, A "marl mesa" at Anderson Basin #1 (Fig. 4). The excavations expose Bw-71 (Fig. 23). Stratum 3c forms the resistant cap. The carbonate is underlain by sand, in turn underlain by stratum 2m, barely visible as the dark zone low in the section (arrow) and forming a resistant layer just above the surface the people stand on. D, West wall of the Wind Pit at the Lubbock Landfill (Figs. 4, 27). Figure stands at section W-4, showing the channel within stratum 2. Stratum 2d at left is almost pure diatomite. Note organic staining within stratum 1 and the beds high in organic-matter at the base of the channel. The thick A horizon in stratum 3c also is apparent. E, Archaeological excavations at Plant X (Fig. 4; the Henry's Beach site of Honea, 1980), looking east. The site is in a blowout on the floor of Blackwater Draw where the Muleshoe Dunes cross the drainage. The white sediment exposed around the crew members in the center of the photo is stratum 3c, revealed by deflation of the dunes (consisting largely of stratum 4s). The general appearance of the setting and in particular the exposure of 3c is typical of Blackwater Draw where it passes through the dune fields. F, Exposure of the overthickened variant of the Lubbock Lake soil at the Lubbock Lake site in lower Yellowhouse Draw (Fig. 4). The B horizon and lower A horizon formed in stratum 4s. Most of the A horizon formed as stratum 5m slowly accumulated on the floor of the draw.

A

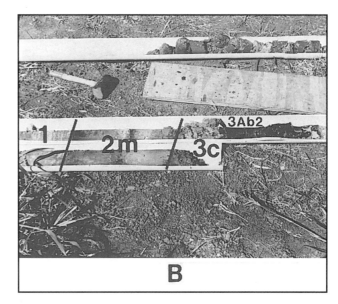

B

C

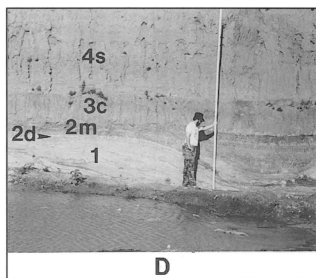

D

E

F

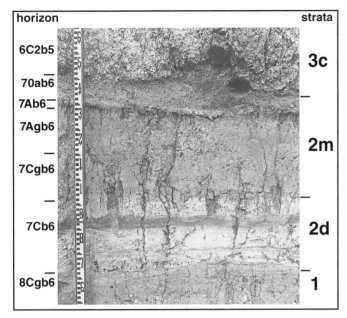

Figure 21. An exposure of stratum 2 at Lubbock Lake (Fig. 4) with an excellent example of the bedded diatomite of stratum 2d and the valley-axis facies of the soil formed in stratum 2m ("horizon" is the soil horizon sequence and "strata" is the lithostratigraphy; rod numbered in decimeters and graduated in centimeters; modified from Holliday, 1985c, fig. 4).

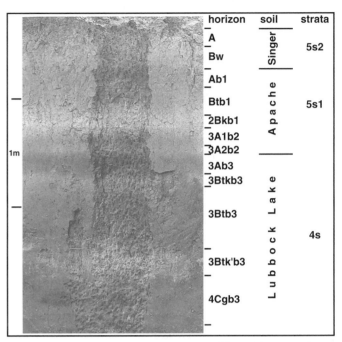

Figure 22. Strata 4s and 5s at Lubbock Lake with excellent examples of the morphology of the buried Lubbock Lake soil and the soils of stratum 5 ("horizon" is the generalized soil horizon sequence, "soil" is the pedostratigraphy and "strata" is the lithostratigraphy; modified from Holliday, 1985c, fig. 5). The Apache and Singer soils are local informal pedostratigraphic units.

TABLE 7A. CORRELATION OF STRATIGRAPHIC TERMINOLOGY FOR THE PLAINVIEW SITE, RUNNING WATER DRAW*

Sellards et al. 1947	Holliday 1985b	Holliday 1990b	Holliday This Volume
Dark, sandy soil	Unit 5	Unit 5B	Stratum 5B
		Unit 5A	Stratum 5A
Massive, compact sandy clay	Unit 4	Unit 4	Stratum 4s
n.r.	n.r.	Unit 3	Strata 3c and 3s
n.r.	n.r.	Unit 2 (top)	Stratum 2m
Basal sand and gravel	Unit 3 (sand)	Unit 2	Stratum 1
	Unit 1 (gravel)		
n.r.	n.r.	Unit 1	Loamy mud filling bedrock channel
Panhandle Formation	Calcrete bedrock		Ogallala Formation Caprock Caliche

*Figures 4, 26.
n.r. = strata not recognized.

**TABLE 7B. ARCHAEOLOGICAL TRADITIONS AND ASSOCIATED (SELECTED) TERRESTRIAL
VERTEBRATE FAUNAS REPORTED FROM RUNNING WATER DRAW**

Stratum	Archaeological Tradition	Vertebrate Fauna	Sites
5	None	*Bison bison*	Quincy Street
1	Plainview	*Bison antiquus*	Plainview*
	None	*Equus* sp.	Plains Paving†

*Sellards et al., 1947; Speer, 1990.
†Hughes and Guffee, 1976.

**TABLE 7C. LITHOLOGIC AND PEDOLOGIC CHARACTERISTICS
OF RUNNING WATER DRAW**

Stratum/ Soil	Range in Thickness (cm)	Description
Unnamed soil		A (ochric or mollic) - Bw; or A (ochric or mollic) - Bw/Bk (Stage I calcic).
5	23 - 100 59 - 130	5s: SL; SiL locally 5m: CL
Lubbock Lake soil		**Valley-axis:** A (mollic) - Bt/Btk (argillic; Stage I-II calcic); A is locally comulic and clayey due to slow deposition of stratum 5m. **Valley-margin:** A (ochric) - Bt/Btk (wk argillic; Stage I-II calcic).
4	50 - 362	4s: SL, CL; clayey locally.
Yellowhouse soil		3c: A (ochric) - C 3s: A (ochric) - Btk (argillic?) or A (ochric) - Bk
3	9 - 113 160 - 200	3c: SCL-SL-SC; 9-48% carb 3s: L 3m (Plainview Landfill): discontinuous, contorted interbeds of S and OM-rich C (5-25 cm).
2	....	2d (Flagg): calcareous diatomite (9 cm) over interbedded diatomite and OM-rich SiC (13 cm) over interbedded S and gleyed, OM-rich C (101 cm). 2m (Edmonson): OM-rich C (5 cm) over interbedded gleyed, OM-rich C and gleyed S (21 cm). 2m (Plains Paving and Plainview): discontinuous, contorted interbeds of S and OM-rich C (5-25 cm).
1	14 - >200	Massive fine sand (w/local cross-beds or contorted bedding) over m-c gr.

See notes with Table 6D for lithologic and pedologic descriptions.

margin facies of stratum 1 were not recognized at other sites, although data are inadequate on the geometry of individual strata at most other sites.

The terraces affording exposures of stratum 1 provide evidence of episodic downcutting prior to final incision of some draws. At the Plainview site an organic-rich mud fills a channel 1 m deep, cut in a bench (strath terrace?) of strongly silicified Caprock calcrete on the south side of the draw (Fig. 20A). The surface of the bench is approximately 1 m higher than the deepest point of valley incision into the calcrete. Similarly, a terrace cut on the Blackwater Draw Formation was observed in trenches on the south side of the draw at Quincy Street, 1 km above the Plainview site (Fig. 26A). This terrace is 1.5 m higher than the floor of the channel. Stratum 1 was deposited on the terrace at Quincy Street and across the bedrock floor of the valley incised below the bench at both sites (Fig. 26B).

At the Lubbock Landfill stratum 1 occurs on two terraces (Fig. 27) and at Lubbock Lake the deposit is preserved on three terraces, one of which is buried (Fig. 28). All terraces at both sites are cut into indurated lacustrine carbonate (Blanco Formation). The two terraces at the Landfill and the two highest terraces at Lubbock Lake are covered by thin layers of Holocene eolian sediment. On the lowest terrace at Lubbock Lake, the sand and gravel (stratum 03 of Stafford, 1981) are up to 1 m thick and buried by valley fill (Holliday, 1985c; Fig. 28B). Johnson and Stafford (1976) also report a buried terrace downdraw from Lubbock Lake.

The age of initial stratum 1 deposition is poorly known, but upper stratum 1 is relatively well dated (Appendix 2). Deposition ended as early as 11,300 yr B.P. in some reaches and continued until as late as 8,900 yr B.P. in other reaches, but more typically, deposition ceased between 11,000 and 9,500 yr B.P. (Table 3).

Stratum 2

Stratum 2 is one of the best known of late Quaternary deposits on the Southern High Plains, owing to the rich archaeological and paleontological associations at the Clovis, Anderson Basin, and Lubbock Lake sites (e.g., Cotter, 1937; Howard, 1935a, b; Stock and Bode, 1936; Antevs, 1949; Sellards, 1952;

Haynes and Agogino, 1966; Holliday, 1985c; Johnson, 1987d). For all of its fame, stratum 2 is quite rare relative to the other strata. Of the 110 study localities, only 16 contain stratum 2. Most are in Running Water (4 sites) and Blackwater Draws (8 sites), and there are 2 each in Yellowhouse and Mustang Draws (Tables 6D, 7C, 9C, 11C). Frederick (1994) also reports a stratum 2 equivalent in lower Sulphur Springs Draw.

Stratum 2 contains beds of diatomaceous earth and pure diatomite (stratum 2d) and noncalcareous or low-carbonate mud (stratum 2m; Fig. 21), both associated with extinct vertebrates and Paleoindian archaeological materials. Stratum 2d typically is composed of layers of diatomite and diatomaceous mud interbedded with mud (Fig. 21). The mud in strata 2d and 2m is

dark gray to black or locally dark olive gray, and silty to clayey. Sandy interbeds in 2m are locally common. Stafford (1981) also reported a carbonate facies of stratum 2 at Lubbock Lake, interpreted as resulting from spring discharge. Stratum 2 is usually 50–200 cm thick and occurs along the valley axes. A sandy, valley-margin facies of stratum 2 (2s) is relatively common (Fig. 6). In lower Sulphur Springs Draw stratum 2 also contains secondary gypsum (Frederick, 1994).

Long or numerous exposures and cores at the Lubbock Landfill, Lubbock Lake, Gibson, and Mustang Springs sites yielded abundant evidence for the geometry and microstratigraphy of stratum 2. At Lubbock Lake and the Gibson site the deposit occupies roughly the width of the draw (Figs. 28B, 29B) and occurs as one or several large bodies that extend for several kilometers along the draw. At the Lubbock Landfill and Mustang Springs (Meltzer, 1991) stratum 2 is relatively localized in a narrow channel inset in stratum 1 (Figs. 20D, 27B, 30B). The Lubbock Landfill exposed a channel in stratum 2, illustrating a complex record of cutting and filling in lower 2d (Figs. 20D, 27B). Three sand layers occur in and adjacent to the channel, interbedded with five layers of 2d and 2m. At Lubbock Lake a small channel (100 cm wide) filled with interbedded mud, sand, and gravel, and choked with *Bison antiquus* bone was discovered near the base of 2d, denoting a final stage of alluviation early in the record of diatomite deposition.

Evidence for pedogenesis in upper stratum 2 occurs locally. The top of stratum 2 at Lubbock Lake has a weakly developed soil (the Firstview soil of Holliday, 1985c, e; Fig. 21). Along the valley axis the soil exhibits A-C and A-Cg profiles and silicified plants remains are ubiquitous. Along the better-drained valley margins the soil has an A-Bw profile. At Clovis and Anderson

TABLE 8A. CORRELATION OF STRATIGRAPHIC TERMINOLOGY FOR THE GIBSON SITE,* MIDDLE BLACKWATER DRAW†

Honea 1980	Holliday This Volume
Zone V	Stratum 5s
Zone IV Zone IIIB Zone IIIA	Stratum 3s
Zone II	Stratum 2s
	Stratum 2d
Zone I	Stratum 1

*Marks Beach site of Honea, 1980.
†Figures 4, 29.

TABLE 8B. ARCHAEOLOGICAL TRADITIONS AND ASSOCIATED TERRESTRIAL VERTEBRATE FAUNAS REPORTED FROM MIDDLE BLACKWATER DRAW* AND LOWER BLACKWATER DRAW†

Stratum	Archaeological Tradition	Vertebrate Fauna	Sites
5s	Late Prehistoric	*Bison bison*	Gibson Ranch[§]
3s	Archaic	*Bison* sp., *Antilocapra americana*	Gibson Ranch[§]
2m, 2d	Paleoindian	*Bison antiquus*	Gibson Ranch[§] Lubbock Landfill[**]
1	None	*Mammuthus columbi*	Plant X[‡]
	None	*Mammuthus columbi*, *Camelops* sp., *Equus* sp.	BFI[§§]

*Gibson and Plant X area, Figure 4.
†BFI and Lubbock Landfill area, Figure 4.
[§]Honea, 1980.
**Brown et al., 1993.
[‡]Honea, 1980, addendum p. 329.
[§§]Holliday, 1983.

Basin #2, stratum 2s (the "carbonaceous silt" of Sellards, 1952; Sellards and Evans, 1960; and Hester, 1975; Figs. 23, 25; Table 6A) represents the accumulations of both silt and sand in an aggrading, organic-rich, probably marshy environment. The layer, therefore, also is a cumulic A horizon (the "wet meadow soil" of Haynes, 1975, Table 4-2). Weak pedogenic modification of sandy, basin-margin facies of stratum 2 also is noted at Clovis (Haynes, 1995). Soils in the sandy facies of stratum 2 serve to stratigraphically separate this sand from overlying basin- or valley-margin sands of stratum 3.

Stratum 2 is very well dated because of its archaeological significance (Table 3; Appendix 2). Accumulation of the lacustrine and palustrine sediments began by ca. 10,000 yr B.P. at many sites and as early as ca. 11,000 yr B.P. at several localities. Deposition of stratum 2 ended before 9,000 yr B.P. at most sites (Table 3). There are three notable exceptions where stratum 2 continued accumulating well into the Holocene: at Clovis and Lubbock Lake, where deposition continued until at least 8,500 yr B.P., and Mustang Springs, where stratum 2 sedimentation continued until at least 8,100 yr B.P.

Stratum 3

Above strata 1 and 2 is a deposit high in calcium carbonate (20–50%), stratum 3c, typically 50–100 cm thick. Stratum 3c is ubiquitous along the valley axes of the draws. This layer is massive and friable when dry. The carbonate content gives the layer a silty to clayey feel, but significant amounts of sand are usually present; abundant silt and clay occur locally.

In most reaches of the draws of the Colorado drainage stratum 3c is distinctly coarser, lower in carbonate, and not as friable as the same deposit in the draws to the north. These characteristics probably are due to the sandy nature of the surrounding Blackwater Draw Formation. At a few sites along lower Mustang Draw, Monahans Draw, and Midland Draw, however, 3c is identical to its correlate in the Brazos system draws. In Upper Mustang Draw stratum 3c sometimes was difficult to differentiate from the calcic horizon of the soil formed in stratum 4. These zones usually could be distinguished based on distinct textural differences between the two layers and on carbonate morphology (the Bk horizon of the soil in stratum 4 consisted of films and threads of carbonate whereas, the carbonate in 3c was very finely divided throughout the layer).

Stratum 3 locally has a low-carbonate, valley margin facies (3s) with sandy to loamy textures. This facies is most common in the draws of the Brazos system and in lower Mustang Draw. In Blackwater Draw, 3s is present along the valley margins in the upper half of the drainage where the draw is located near or in the Muleshoe Dunes (Fig. 29B). At the Clovis site 3s occurs as a sheet sand draped across the ancient basin (Fig. 25) and probably has a dunal facies on the uplands surrounding the basin. On lower Blackwater, 3s is a loamy deposit along the valley margins at the Lubbock Landfill and BFI sites and on top of the strath terraces at the landfill (Fig. 27B). At these sites and at the Midland site in Monahans Draw, the layer can be subdivided

TABLE 9A. CORRELATION OF SELECTED STRATIGRAPHIC TERMINOLOGY FOR THE LUBBOCK LAKE SITE, LOWER YELLOWHOUSE DRAW*

Stafford 1981	Holliday 1985c	Holliday This Volume
Stratum 5C	Strata 4Bl, 5Al, and 5Bl	Strata 4m and 5m
Stratum 5B Stratum 5A	Stratum 5B Stratum 5A	Strata 5s2 and 5g2 Strata 5s1 and 5g1
Stratum 4C	A-horizon	
Stratum 4B	Stratum 4B	Statum 4s
Streatum 4A		Stratum 4s
Stratum 3C clay	Stratum 4A	Stratum 4m
Stratum 3C A-horizon	A-horizon	
Stratum 3B Stratum 3A	Stratum 3l Stratum 3e	Stratum 3m Stratum 3s
Stratum 2C	A-horizon	Stratum 2s or 2m
Stratum 2E	Stratum 2s	Stratum 2s
Stratum 2F	Stratum 2e and 2F	
Stratum 2B Stratum 2A	Stratum 2B Stratum 2A	Stratum 2m Stratum 2d
Stratum 1C Stratum 1B Stratum 1A	Stratum 1	Stratum 1

*Figures 4, 28. Modified from Johnson, 1987b, Table 1.1.

into 3s1 and 3s2. At Midland, stratum 3s2 yielded well-known human skeletal remains (Wendorf et al., 1955; Table 13B). Slight increase in organic carbon content in 3s2 relative to strata 3s1 and 4s indicate that this layer may have been an A horizon. A highly localized, sandy and gypsiferous facies of stratum 3 also is reported below the marl facies in lower Sulphur Springs Draw (Frederick, 1994).

Stratum 3 is the surface deposit in some draws. In Blackwater Draw the layer was removed by deflation above Anderson Basin #1, but commonly is exposed along the floor of the draw in the area of the Baker site and beginning approximately 10 km east of the Muleshoe site and continuing down draw to the area of the McNeese site (Figs. 4, 20E). At Anderson Basin #1 and #2, where most of the valley fill was removed by deflation, stratum 3c locally forms a resistant layer resulting in development of small "mesas" (Figs. 20C, 23). Stratum 3 also is the surface deposit in some reaches of Seminole, McKenzie, and lower Mustang Draws.

At the top of stratum 3 at many sites is the A horizon of a

TABLE 9B. ARCHAEOLOGICAL TRADITIONS AND ASSOCIATED (SELECTED) TERRESTRIAL VERTEBRATE FAUNAS REPORTED FROM LOWER YELLOWHOUSE DRAW

Stratum	Archaeological Tradition	Vertebrate Fauna	Sites
5	Historic	*Bison bison, Equus caballus, Antilocapra americana*	Lubbock Lake and Lower Yellowhouse Draw*
	Late Prehistoric	*Bison bison, Antilocapra americana*	Lubbock Lake and Lower Yellowhouse Draw*
4	Late Prehistoric	*Bison bison, Antilocapra americana*	Lubbock Lake†
	Archaic	*Bison bison, Antilocapra americana*	Lubbock Lake†
2s	Lake Paleoindian	*Bison antiquus, Antilocapra americana*	Lubbock Lake and Lower Yellowhouse Draw*
2d	Folsom	*Bison antiquus, Capromeryx sp.*	Lubbock Lake and Lower Yellowhouse Draw*
1	Clovis (probable)	*Geochelone, Mammuthus columbi, Camelops hesternus, Equus mexicanus and E. Francisi, Bison antiquus, Arctodus simus, Holmesina septentrionale*	Lubbock Lake and Lower Yellowhouse Draw*

*Holliday, 1985c; Johnson, 1986, 1987c, d; Johnson and Stafford, 1976.
†Holliday, 1985c; Johnson, 1986, 1987c, d.

weakly developed soil (the Yellowhouse soil; Fig. 6). The A horizon typically is dark gray to black, but otherwise this soil has no distinguishing characteristics and the C horizon is unaltered stratum 3. Along the valley axes, the A horizon locally is high in clay, taking on the characteristics of the older muds of stratum 2 and referred to as stratum 3m (Fig. 6). A weakly developed, buried A horizon also occurs locally within 3c. The Yellowhouse soil in stratum 3s has a thinner and sandier A horizon, and a sandy, oxidized, low-carbonate C horizon. At the Lubbock Landfill, most of 3c is represented by an overthickened A horizon (Figs. 20D, 27B) with physical characteristics (low bulk density, low carbonate content, relatively high organic carbon content) similar to stratum 2. The A horizon in 3s along the valley axis at the Landfill also is overthickened, suggesting eolian deposition in a somewhat damp, slowly aggrading setting.

There are more radiocarbon ages for stratum 3 than any other late Quaternary deposit on the Southern High Plains (Appendix 2). The layer is time-transgressive along the draws, accumulating largely in the early Holocene, between 10,000 and 7,500 yr B.P. (Table 3). Locally, deposition of stratum 3 continued into the middle Holocene, but dates of <6,500 yr B.P. are from A-horizons and denote burial by stratum 4s or 5s rather than accumulation of 3c or 3s.

Stratum 4

The youngest widespread deposit in the draws is stratum 4, usually resting conformably on stratum 3. Unconformable contacts are noted only at Clovis (Haynes, 1995), Lubbock Lake (Stafford, 1981; Holliday, 1985c), and Mustang Springs (Meltzer, 1991). Stratum 4 is 1–3 m thick and generally consists of a massive, sandy to loamy and, rarely, clayey deposit (all identified as facies 4s) 1–3 m thick. Silty (4si) and muddy (4m) facies occur in some sites.

Facies variations in stratum 4 are relatively rare. The most common is a coarsening in texture toward the valley margins, which usually is not significant enough for separate identification. Also common is a sandy dunal facies. At the Clovis site, stratum 4 occurs as a sheet sand across the ancient basin and as dunes on the uplands surrounding the basin (Fig. 25). Below the site, stratum 4 is found in dunes on the floor of the draw at Anderson Basin #1 and #2. Stratum 4 also occurs as dunes in the draws and draped across uplands at Plant X and the Gibson site (middle Blackwater Draw; Fig. 20E), between Milnesand and Bledsoe (upper Sulphur Draw), and at the Midland site (Monahans Draw; Fig. 24B). Locally the lower portion of stratum 4 is clayey and dark, apparently representing cumulization of the Yel-

TABLE 9C. LITHOLOGIC AND PEDOLOGIC CHARACTERISTICS OF YELLOWHOUSE DRAW

Stratum/ Soil	Range in Thickness (cm)	Description
Unnamed soil		A (ochric or mollic) - Bw; or A (ochric or mollic) - Bw/Bk (Stage I calcic).
5	10 - 75 34 - 100	**5s and 5g undivided**: SCL, SL, and bedded gr. **5m**: L, CL SL, SiL.
Lubbock Lake soil		**Valley-axis**: A (mollic) - Bt/Btk (argillic; Stage I-II calcic); A is locally cumulic due to slow deposition of stratum 5m. **Valley-margin**: A (ochric) - Bt/Btk (wk argillic; Stage I-II calcic).
4	50 - 300*	**4s**: SCL, SL; L and CL locally; sandy valley-margin facies. **4m**: C; valley axis only.
Yellowhouse soil		**3c**: A (ochric, locally cumulic) - C
3	30 - 120 48 - 155	**3c**: SiL, SiCL, CL, SCL, SL, LS; 17-57% carb; local lenses of low carbonate sand. **3s**: SL
2	....	**2s**: (Lubbock Lake): SCL, SL; valley margin only. **2m** (Lubbock Lake): SiC with silicified plant remains common (30-80 cm).† **2d** (Lubbock Lake): interbedded diatomite and SiC (30-100 cm); local lenses of fS common.†
1	0 - >200	Massive, fine, quartz sand (with local cross-beds or contorted bedding) over m-c, locally bedded gr; interbeds of clay and loamy sand (1-10 mm) and a clay drape (10 cm) locally common at and below Lubbock Lake.

*<150 cm above the Narrows (except Claunch, the one site on the North Fork); 120-250 cm thick below the Narrows (see Fig. 4 for locations).

†Identical deposits reported in Yellowhouse Draw below Lubbock Lake (Johnson and Stafford, 1976; Stafford, 1981).

See notes with Table 6D for lithologic and pedologic descriptions.

TABLE 10. LITHOLOGIC AND PEDOLOGIC CHARACTERISTICS OF SULPHUR DRAW

Stratum/ Soil	Range in Thickness (cm)	Description
Lubbock Lake soil		**Valley-axis**: A (ochric, locally mollic) - Bt/Btk (argillic; Stage I-II calcic). **Valley-margin**: A (ochric) - Bt/Btk (weak argillic; Stage I-II calcic).
4	120 - 310	**4s**: SL, LS.
Yellowhouse soil		**3c**: A (ochric) - C
3	10 - 120	**3c**: SL, SCL; 16-33% carb.
1	0 - 100	Massive, fine, quartz sand; interbedded with or over m-c gr.

See notes with Table 6D for lithologic and pedologic descriptions.

TABLE 11A. CORRELATION OF STRATIGRAPHIC TERMINOLOGY FOR THE MUSTANG SPRINGS SITE, LOWER MUSTANG DRAW*

Meltzer and Collins 1987	Meltzer 1991	Holliday This Volume
Stratum 7 Stratum 6 Stratum 5	Stratum 5	Stratum 5s Stratum 5s Stratum 5m
Stratum 4	Stratum 4	Stratum 4s
Stratum 3H	Stratum 3	Stratum 3c
Strata 3A-3G	Stratum 2	Strata 2d and 2m
Stratum 2 Stratum 1	Stratum 1	Stratum 1

*Figures 4, 30.

lowhouse soil as stratum 4 deposition began. At the Lubbock Landfill stratum 4s is composed of two layers (4s1 and 4s2; Fig. 27B). At and below Lubbock Lake a clayey mud (4m) occurs as a lateral, valley-axis facies of 4s (Fig. 28B).

The most distinctive physical characteristics of stratum 4 are the result of pedogenesis. A moderately to strongly developed soil (the Lubbock Lake soil) formed in the layer (Figs. 20F, 22). This soil usually is the modern surface soil along the draws. The valley-axis profiles typically are more strongly developed (mollic A—argillic and calcic Bt/Bk) than the valley margin profiles (ochric A—cambic Bw or Bk). Along valley axes, some profiles also are cumulic, with over-thickened A horizons resulting from

slow accumulation of clayey sediment (stratum 5m, discussed below) and organic matter on top of stratum 4 after pedogenesis began (Fig. 20F). In the dunal facies of the Lubbock Lake soil at Clovis, Gibson, and Midland illuvial clay occurs as "clay bands" like those described by Gile (1979) in the nearby Muleshoe Dunes. At the Lubbock Landfill a weakly developed soil (A-C profile) also formed in stratum 4s1 (Fig. 27B).

Stratum 4 is discontinuous in some draws. At Barwise in lower Running Water Draw the deposit probably was removed by headcutting. Barwise is only 5 km above the upper end of Blanco Canyon where the drainage is undergoing deep incision. Stratum 4 is missing along much of upper Blackwater Draw. The layer is not preserved along the draw adjacent to and above the

TABLE 11B. ARCHAEOLOGICAL TRADITIONS AND ASSOCIATED (SELECTED) TERRESTRIAL VERTEBRATE FAUNAS REPORTED FROM LOWER MUSTANG DRAW

Stratum	Archaeological Tradition	Vertebrate Fauna	Sites
5	None	*Bison bison*	Mustang Springs*
4	Archaic	None	Mustang Springs*
2	None	*Bison antiquus?*	Mustang Springs*
1	None	*Mummuthus columbi*	Evans and Meade†

*Meltzer, 1991.
†Evans and Meade, 1945 (locality 32 = Evans and Meade site).

TABLE 11C. LITHOLOGIC AND PEDOLOGIC CHARACTERISTICS OF LOWER MUSTANG DRAW

Stratum/ Soil	Range in Thickness (cm)	Description
Lubbock Lake soil		A (ochric) - Bw or Bk (cambic and Stage I-II calcic); locally Bt/Btk argillic and calcic.
4	57 - 185	**4s**: L, SL; silty locally.
Yellowhouse soil		**3c**: A (ochric, locally cumulic) - C
3	25 - 240	**3c**: SiL, CL, L, SCL; 16-61% carb.
2	....	**2m** (Wroe Ranch): mud (20 cm); secondary silica common. **2d** (Mustang Springs): interbedded diatomite and SiC (100 cm); local lenses of fS common.
1	0 - 400	Massive, fine, quartz sand (w/local cross-beds or contorted bedding); local interbeds of clay and loam; m-c gr locally common below sand.

See notes with Table 6D for lithologic and pedologic descriptions.

TABLE 12. LITHOLOGIC AND PEDOLOGIC CHARACTERISTICS OF MIDLAND DRAW

Stratum/ Soil	Range in Thickness (cm)	Description
Lubbock Lake soil		A (ochric) - Bt-Btk (argillic and Stage I-II calcic).
4	80 - 174	**4s**: SCL, L, CL
Yellowhouse soil		**3c**: A (ochric) - C
3	76 - 200	**3c**: SL; 17-47% carb; OM-rich interbeds locally common.
1	0 - >30	LfS or fSL, quartz sand.

See notes with Table 6D for lithologic and pedologic descriptions.

TABLE 13A. CORRELATION OF STRATIGRAPHIC TERMINOLOGY FOR THE MIDLAND SITE, MONAHANS DRAW*

	Wendorf et al. 1955		Holliday This Volume
	Monahans Formation	Unit 5	Stratum 5s
		Unit 4	
Judkins Formation	Red sand	Unit 3	Stratum 4s
	Gray calcareous sand	Unit 2b	Stratum 3s2
		Unit 2a	Stratum 3s1
	White calcareous sand	Unit 1	Stratum 1

*Figures 4, 24.

Clovis site or at the Baker site, but it does occur as an essentially continuous sheet from the Davis to the Birdwell sites. Between the Birdwell and McNeese sites, localized deposits of stratum 4 were found at the Muleshoe, Halsell, Plant X, and Gibson sites. Below McNeese to the mouth of the draw, stratum 4 is a continuous deposit. Stratum 4 was not observed in Yellowhouse Draw at the lower end of the Narrows and in a few reaches of Seminole, McKenzie, and lower Mustang Draws.

Dating stratum 4 deposition is based on radiocarbon ages from the top of stratum 3, discussed above, and ages from within stratum 4. The formation of the Lubbock Lake soil is dated on the basis of radiocarbon ages from the top of the A horizon (noncumulic facies) and ages from overlying stratum 5 (Appendix 2). Much of the age control is published elsewhere (Appendix 2). Stratum 4 was deposited largely in the middle Holocene,

TABLE 13B. ARCHAEOLOGICAL TRADITIONS AND ASSOCIATED (SELECTED) TERRESTRIAL VERTEBRATE FAUNAS REPORTED FROM MONAHANS DRAW

Stratum	Archaeological Tradition	Vertebrate Fauna*	Sites
4	None	*Capromeryx* (?), *Equus* sp., *Bison antiquus*	Midland†
3s2	???	*Homo sapiens, Equus* sp., *Capromeryx, Bison antiquus*	Midland†
3s1	None	None	
1	None	*Mammuthus columbi, Camelops* sp., *Equus* sp., *Canis dirus, Capromeryx, Bison antiquus*	Midland†

*All remains of extinct species from strata 3 and 4 are fragmentary.
†Wendorf et al., 1955; Wendorf and Krieger, 1959; Sellards, 1955b.

TABLE 13C. LITHOLOGIC AND PEDOLOGIC CHARACTERISTICS OF MONAHANS DRAW ABOVE THE MIDLAND SITE*

Stratum/ Soil	Range in Thickness (cm)	Description
Lubbock Lake soil		A (ochric) -Bt (argillic)
4	155	**4s:** L, SCL
3	128	**3c:** SL-LS; 23-26% carb.

*Core Mn-18, Figure 24A.
See notes with Table 6D for lithologic and pedologic descriptions.

TABLE 13D. LITHOLOGIC AND PEDOLOGIC CHARACTERISTICS OF THE MIDLAND SITE, MONAHANS DRAW

Stratum/ Soil	Range in Thickness (cm)	Description
Unnamed soil		A (ochric) - C
5	0 - 100	**5s:** S
Lubbock Lake soil		A (Ochric) - Bw and Bt (Bw with thin zone of clay bands present).
4	0 - 100	**4s:** S, LS
3	20 - 80+	**3s2:** S, LS, SL; 0-14% carb. **3s1:** S
1	60+	S, LS, SL; 0-25% carb.

See notes with Table 6D for lithologic and pedologic descriptions.

between 7,500 and 4,500 yr B.P. Some radiocarbon ages from stratum 4s overlap with ages on strata 2 and 3 (Table 3), a result of local difficulties in stratigraphic separation noted above. Pedogenic alteration of stratum 4 followed throughout the late Holocene (Table 3). The Lubbock Lake soil is the surface soil on the floors of the draws where stratum 4 was deposited, but is not buried by stratum 5.

Stratum 5

There are several types of localized deposits above the much more common stratum 4, and all are identified as facies of stratum 5. These deposits are most common, although discontinuous, along Running Water, Blackwater, Yellowhouse, and lower Mustang Draws. The most common facies is an organic-rich, muddy layer (5m) found along the axes of some draws. Stratum 5m usually is welded to the top of the A horizon of the Lubbock Lake soil, producing a cumulic A horizon (Fig. 20F). Within the cumulized A horizon, stratum 5 is differentiated from stratum 4 based on texture; the former having less sand than the latter. Stratum 5m is common along middle Running Water Draw and is ubiquitous along the axis of lower Yellowhouse beginning in the area of the County Caliche Pit and continuing below the confluence with Blackwater (Fig. 20F). A likely equivalent of stratum 5m was found in lower Sulphur Springs Draw (stratum 6; Fredrick, 1994). In lower Yellowhouse and lower Sulphur Springs Draws, 5m locally is inset into older deposits (Fig. 28B).

There are localized accumulations of silty, sandy, and sandy gravel facies (5s), and gravelly facies (5g) of stratum 5 along both the valley axes and valley margins. The most extensive accumulations are known from Plainview (Running Water Draw), Lubbock Landfill (Blackwater Draw), and Lubbock Lake (Yellowhouse Draw), where large exposures are accessible (Figs. 26B, 27B, 28B). At all sites strata 5s and 5g occur along the valley margins and can be traced into stratum 5m. Stratum 5s also occurs as two layers (5s1 and 5s2) at Plainview, Evans, and Lubbock Lake. Weakly to moderately well developed soils formed in strata 5s and 5g and exhibit A (ochric)–Bw (cambic) or A

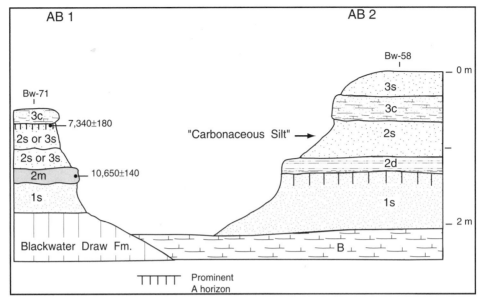

Figure 23. Stratigraphic sections at Anderson Basin (Figs. 4, 10A) with locations of radiocarbon samples. Anderson Basin #1 (AB1) is to the left and Anderson Basin #2 (AB2) is to the right. The horizontal relationship is schematic and not to scale; the sections are ca. 2.5 km apart. See Figure 6 for key to symbols.

(ochric)–Bk/Bt/Btk (argillic and/or calcic) profiles. At Lubbock Lake 5s1 and 5s2 correspond to the Apache and Singer soils, respectively (Holliday, 1985c, d; Fig. 22). These soil names are not used beyond Lubbock Lake because of the highly localized distribution of strata 5s and 5g.

Several less common facies of stratum 5 were found at a few sites. A massive, primary carbonate formed at the Muleshoe site in Blackwater Draw and at the Payne and Lubbock Lake sites in Yellowhouse Draw. Stratum 5s occurs in dunes in the draws and on uplands at the Gibson site and in the Clovis area (Tables 6A, 6D, 8A; Fig. 29B) in Blackwater Draw, and at the Midland site in Monahans Draw.

Evidence for erosion preceding and during stratum 5 deposition was found at several sites. At Quincy, in Running Water Draw, a deep channel (>300 cm) was cut into stratum 4 and filled with layers of gravel, sand, loam, mud, and carbonate. This channel contains the only evidence of late Holocene alluviation documented in any of the draws examined as part of this study. In Yellowhouse Draw at and below Lubbock Lake considerable downcutting occurred along the valley axis before and probably during deposition of stratum 5m. Locally, stratum 5m is inset against strata 4 and 3c (Fig. 28B). Evidence for cutting and filling within 5m is not visually apparent due to the homogenous nature of the deposit, but is indicated by the wide range of radiocarbon ages from the deposit and the age variation (based on radiocarbon ages, artifact assemblages, and faunal types) among archaeological features contained in the unit (Johnson, 1987b). In Mustang Draw at and below Mustang Springs stratum 5m rests conformably on stratum 4 and unconformably on strata 1 and 3 where strata 4, 3, and/or 2 are missing (Fig. 30B). Very

recent erosion of 5m at Mustang Springs is indicated by a steep-walled depression up to 1 m deep cut into this layer (Figs. 14E, 30B). Erosion of 5m at Mustang Springs probably is associated with historic spring discharge (Meltzer, 1991).

Stratum 5 is dated on the basis of radiocarbon ages from the A horizons formed in this deposit, from ages on archaeological charcoal, and on ages from the underlying Lubbock Lake soil (Appendix 2). The deposit represents localized deposition and subsequent pedogenesis in the late Holocene, <3,900 yr B.P. (Table 3).

Discussion

Interpretations of the depositional environments of the five valley fill strata are presented in this discussion. Because of differences in the kinds of data available from each strata and from each site, the interpretations are relatively general in some instances and relatively specific in others. Discussions of other aspects of the strata (paleontology, paleobotany, and geochemistry) are presented in succeeding sections.

Stratum 1 primarily is an alluvial channel-axis deposit. This interpretation is based on the occurrence of cross-bedded and well-sorted gravel and sand, localized fining upward sequences, and the frequent occurrence of stratum 1 in terrace positions. Molluscs from stratum 1 at Plainview and Anderson Basin #1 (see following section, "Paleontology, Paleobotany, and Stable Isotopes") and from Lubbock Lake (Pierce, 1987) also support the interpretation of flowing water. The extent, sorting, gradient, local uniformity of thickness and clast size, and lack of evidence for extensive cut-and-fill sequences suggests that stratum 1 was deposited by a competent, perennial, and probably meandering stream. The fine sand beds in stratum 1 often are devoid of primary sedimentary structures. The massive nature of the sand suggests eolian deposition. This interpretation is proposed for some sections of stratum 1 at Lubbock Lake (Holliday, 1985c). Whatever eolian sedimentation occurred probably was localized and of low intensity, because no paleodunes or extensive sheet sands are associated with stratum 1.

The clays in stratum 1, commonly occurring near or at the top of the layer, have several possible origins, including slackwater, overbank, or lacustrine deposition. Stafford (1981) favored the latter interpretation for clay within stratum 1 at Lubbock Lake. Locally at Lubbock Lake, stratum 1 grades upward from fine sand into clay, which is more suggestive of waning flow energy (Holliday, 1985c). The occurrence of layers of gleyed clay interbedded with lenses of sand and palustrine mud at the base of strata 2 or 3 in some other sections (Flagg, Edmonson, Plainview Landfill, Anderson Basin #1, Progress, Tolk) indicates

that the clays are lacustrine and related to the change in depositional environment from stratum 1 to strata 2 or 3. Molluscs from stratum 1 at the Wroe site also suggest the onset of lacustrine conditions (see following section "Paleontology, Paleobotany, and Stable Isotopes").

The diatomite, diatomaceous earth, and mud of stratum 2 are thickest and best expressed along valley axes, and resulted from localized pond and marsh deposition. The beds of pure diatomite were deposited by standing water. The interbeds of mud formed under subaerial conditions, indicated by diatoms (see following section "Paleontology, Paleobotany, and Stable Isotopes") and presence of archaeological features (typically bone beds from bison kills; Johnson, 1987d; Johnson and Holliday, 1980). The interbedded diatomites and muds of stratum 2d, therefore, document fluctuating water levels. The muds of 2m contain terrestrial diatoms, aquatic molluscs, and archaeological features and represent an aggrading marsh with temporary ponds. These and other physical and chemical characteristics of the ponds and marshes are further discussed in the following section, "Paleontology, Paleobotany, and Stable Isotopes."

The stratum 2 environments represent a dramatic hydrologic shift from the flowing water conditions of stratum 1. Radiocarbon ages from Lubbock Lake (Holliday et al., 1983, 1985) indicate that this hydrologic change occurred within several hundred years. Within this period, the depositional environments changed back and forth from alluvial to lacustrine or palustrine conditions, indicated by interbedded sands, clays, and muds at the base of stratum 2 at sites noted above. The shifting depositional envi-

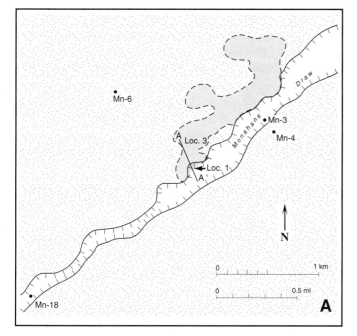

Figure 24. A, Monahans Draw in the area of the Midland site (Fig. 4) showing the locations of the dunes fringing the draw, section A–A′ from Locality 3 in the dunes to Locality 1 in the draw (Fig. 24B), and cores taken outside of the two localities. See Figure 12A for key to symbols. B, Generalized stratigraphic cross section A–A′ across Monahans draw at the Midland site (Fig. 24A) showing locations of Localities 1 and 3w (3 west) and sections Mn-5 and Mn-19. Subdivisions of stratum 3s (3s1 and 3s2) are not illustrated. See Figure 6 for key to symbols.

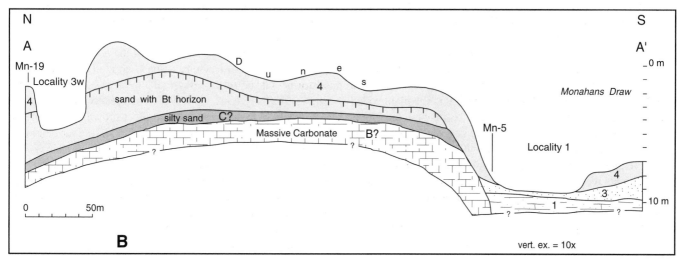

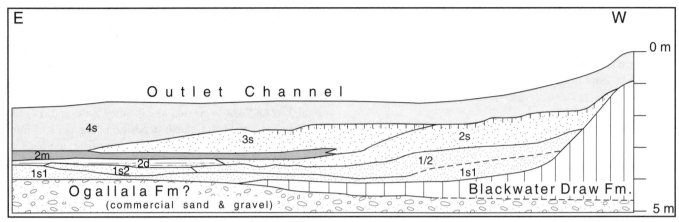

Figure 25. Generalized stratigraphic cross section of the South Bank of the Clovis site (Figs. 4, 12A; based on Stanford et al., 1990, figs. 5, 6; and Haynes, 1995, fig. 7). No horizontal scale. See Figure 6 for key to symbols.

ronments are shown most dramatically in the Wind Pit at the Lubbock Landfill and at Lubbock Lake, where several channels truncate stratum 2, and alluvial sand fills the channels and interfaces with younger layers of stratum 2 (Fig. 20D).

Sandy, valley-margin facies of stratum 2 resulted from either slope wash or eolian additions or, locally, from spring discharge. Layered deposits of sand, locally mixed with coarser sediment, and traceable up valley walls, represent sediments washed into stratum 2. Slope-wash additions are the most common valley-margin facies of stratum 2, occurring at all sites where stratum 2 is preserved. Massive, homogeneous sands, occurring on the upwind (west) sides of valleys in proximity to dunes or traceable into dunes is clearly suggestive of eolian activity during stratum 2 deposition. The eolian facies occurs at Clovis, Anderson Basin, and Gibson in upper Blackwater Draw where it follows the Muleshoe Dunes, and in lower Yellowhouse Draw at Lubbock Lake. Wedge-shaped deposits of well-sorted, convoluted, and sometimes fining-upward layers of sand along basin margins are indicative of spring-laid deposits (Haynes and Agogino, 1966). These sands are best known from the Clovis site (Haynes and Agogino, 1966), but are also present at Lubbock Lake along with spring-laid carbonates (Stafford, 1981). Massive, clean, well-sorted sands, with local, thin interbeds of mud are exposed in blowouts on the south side of Blackwater Draw near the Clovis site (including lower Unit F at the Model T or Car Body locality of Haynes, 1975, p. 90). Artifacts in the sand date the deposits to the Paleoindian period (11,000–8,000 yr B.P.). The interbedded mud and sand is similar to the spring-laid sediment in the Clovis gravel pit and at Lubbock Lake. The massive sands appear to be wind deposited, but derived from an alluvial source such as spring sediments. The lithology and age of these deposits therefore suggests the presence of extensive stratum 2s spring sediments (and spring conduits?) on the opposite side of the draw from the Clovis paleobasin.

The friable, massive, highly calcareous facies of stratum 3 (3c) is one of the most ubiquitous deposits in the draws of the

Llano Estacado. Lithologically identical deposits occur in many of the playa-lake basins of the region (Holliday, 1985a, unpublished data). Moreover, stratum 3c also contains poorly crystallized palygorskite and sepiolite in the clay fraction, based on X-ray diffractometry. These minerals are found in the Pliocene and Pleistocene lacustrine carbonates of the region (Bigham et al., 1980), in the playa-carbonates lithologically similar to 3c, and commonly are associated with lake carbonates in arid and semiarid regions (Singer, 1989). The physical and mineralogical properties of 3c, and its occurrence along the valley axes of the draws suggests that it was deposited by precipitation in marshes or shallow ponds under seasonally warm and dry conditions. The molluscs, diatoms, and ostracodes from 3c (see following section, "Paleontology, Paleobotany, and Stable Isotopes"; Pierce, 1987) support this interpretation. This palustrine facies of stratum 3 is referred to as a marl owing to its lithology and genesis, although it tends to be sandier than some definitions of that term would allow (cf., Bates and Jackson, 1980). The sandier, low-carbonate valley-margin facies probably is of eolian origin based on its homogenous, fine-grained lithology and usual occurrence on the upwind (west) side of the draws. The sand in the marl probably is eolian as well, deposited along the valley-axis as the carbonate accumulated.

The high organic-matter content of the A horizon at the top of 3c and the lack of leaching in the C horizon suggest that the soil formed at the surface of an alkaline marsh, with the water table just below ground level. A high and relatively stable water table would both support luxuriant plant growth at the surface (and promote development of a thick, dark A horizon) and prevent the carbonate from being leached (Holliday, 1985e). The soil in 3s is weakly calcareous and usually thinner, sandier, and lower in organic carbon than its 3c equivalent. These pedologic characteristics reflect development of the soil in a better-drained segment of the landscape, such as a slope position with sandier parent material (Holliday, 1985e).

Stratum 4s is considered to be an eolian deposit, primarily, for several reasons. (1) Stratum 4 is ubiquitous and occurs in a blanket-like manner along most of the draws, as if it was draped across all drainages. (2) Stratum 4 locally is a dunal deposit where draws are covered by or associated with a dune field. (3) The deposit is completely devoid of bedding, stratification, or coarse clastic sediment indicative of an alluvial deposit. (4) The texture of stratum 4 is similar to that of proximal exposures of the Blackwater Draw Formation. This relationship is consistent with redeposition of wind-eroded eolian deposits mantling adjacent uplands. An alluvial deposit should include a mixture of textures representing materials derived from the Blackwater Draw Formation and older units exposed in the draw. (5) The mineralogy of stratum 4 is identical to that of the Blackwater Draw Formation, from which it was presumably derived by deflation. A deposit with a significant component of alluvially derived sediment should have a mineralogy at least partly reflective of the sediments exposed along the draws. The lacustrine carbonates exposed in many reaches of the draws have a mineralogy distinctively different from that of the Blackwater Draw Formation (Bigham et al., 1980; Holliday, 1982; McGrath, 1984).

Clayey, organic-rich facies of stratum 4 (4m) are lithologically similar to strata 2m and 5m and probably resulted from marsh deposition. This interpretation is based on the occurrence of the deposit along valley axes, and its fine-grained, homogeneous, carbonaceous lithology.

Stafford (1981, p. 551) identified "a brief regional nondepositional and erosional episode" between strata 3 and 4 in Running Water, Blackwater, and Yellowhouse Draws. This identification was based largely on work at and below Lubbock Lake in lower Yellowhouse Draw with additional data obtained at: three trenches above Lubbock Lake; three trenches, three quarries, and several cores on lower Blackwater Draw; and cores and two quarries on middle Running Water Draw (Stafford, 1981, p. 549). A hiatus in deposition between strata 3 and 4 and concomitant pedogenesis in stratum 3 is clearly in evidence in the valley fill, but the 49 sites studied in these three draws yielded no evidence for regional erosion save that noted in and near Lubbock Lake and Clovis, and also at Mustang Springs on lower Mustang Draw.

The contact between strata 3 and 4 probably does not represent a dramatic shift in depositional environment (lacustrine to eolian). More likely, it is the result of the end of marsh sedimentation (stratum 3c) and a gradual increase in eolian deposition (beginning during stratum 2 deposition, increasing during stratum 3 accumulation, and culminating with stratum 4). This interpretation is suggested by the usually conformable contact between 3 and 4 and the time-transgressive age of the boundary (see later section, "Summary, Discussion, and Conclusions"). Sediments identified as 4s at some localities likely are the result of the same depositional events producing sediment designated as 2s or 3s at other localities, based on radiocarbon dating (Table 3; Fig. 19). Designation as 4s was made in the absence of obvious stratigraphic evidence for a facies relationship with strata 2 or 3.

Stratum 5 formed in a variety of depositional environments. The mud facies of stratum 5, like 2m and 4m, probably accumulated in marshes and ponds that are reported historically

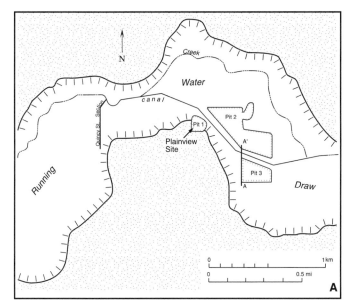

Figure 26. A, Running Water Draw in the area of the Plainview and Quincy Street sites (Fig. 4) with the locations of the three caliche quarries, the original Plainview site quarry (Pit 1), and section A–A′ in Pit 3 (Fig. 26B; modified from Speer, 1990, Fig. 1). See Figure 12A for key to symbols. B, Generalized stratigraphic cross section along the west wall of Pit 3 at the Plainview site (A–A′, Fig. 26A) with locations of stratigraphic sections 2(S2), 3(S3), 7(S7), and 12(S12) (Appendix 1) and all radiocarbon samples (modified from Holliday, 1990b, Fig. 1). See Figure 6 for key to symbols.

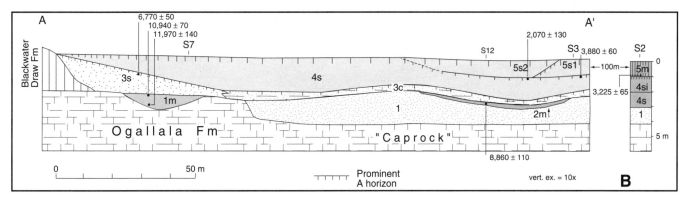

on the floors of some draws (Holden, 1959; Brune, 1981; Bolton, 1990). The gravels and some bedded sands (stratum 5g) are localized slope wash deposits. Stratum 5s, the more massive, unbedded sands and loams with no gravel or other coarse clastics probably are eolian deposits. Localized stream flow (in contrast to slope wash on valley margins) along the axes of Running Water, Yellowhouse, and Mustang Draws is suggested by the lenses of gravel in stratum 5 along the valley at Plainview and in the cut and fill sequences documented at Quincy, Lubbock Lake, and Mustang Springs. Localized erosion of the valley fill occurred during and perhaps before deposition of stratum 5. Alluvium of stratum 5s fills a channel at the Quincy site in Running Water Draw and the marsh deposits of strata 5m fill channels at Lubbock Lake (Yellowhouse Draw) and Mustang Springs (Mustang Draw).

The secondary gypsum reported from the valley fill of lower Sulphur Springs Draw is anomalous relative to the draw sediments at all other sites, although the environments of deposition in the lower reach of the draw were similar to those identified elsewhere. The deposits probably acquired the gypsum by precipitation from ground water. The abundance of saline water at the surface today supports this interpretation. As Frederick (1994, p. 76) points out, the presence of secondary gypsum throughout the younger valley fill "suggests that the water quality in Sulphur Springs Draw has been poor for much of the Holocene."

PALEONTOLOGY, PALEOBOTANY, AND STABLE ISOTOPES

Paleontological and paleobotanical research were significant aspects of many previous studies of draw fills. Vertebrate remains are the best-known fossils in the valleys, but diatoms are the most abundant and other fossils also are known. During the 1988–1993 studies, efforts were made to recover vertebrate fossils, gastropods and bivalves, diatoms, and pollen; and pilot studies were initiated to obtain fossil insects, phytoliths, and ostracodes (see Appendix C of the GSA Data Repository 9541,

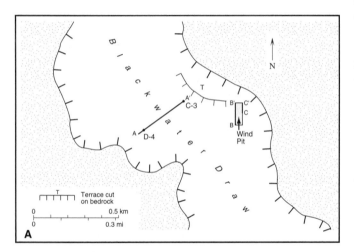

Figure 27. A, Lower Blackwater Draw in the area of the Lubbock Landfill (Fig. 4) with the locations of the terraces cut on bedrock and location of sections A–A′ across the valley (Fig. 27B), and B–B′ (Fig. 27B) and C–C′ (Fig. 11) in the Wind Pit along the valley margin. See Figure 12A for key to symbols. B, Stratigraphic cross sections A–A′ across Blackwater Draw at the Lubbock Landfill and B–B′ on the west wall of the Wind Pit along the valley margin (Fig. 27A), with locations of sections C3, D4, W2, and W4, and all radiocarbon samples. See Figure 6 for key to symbols.

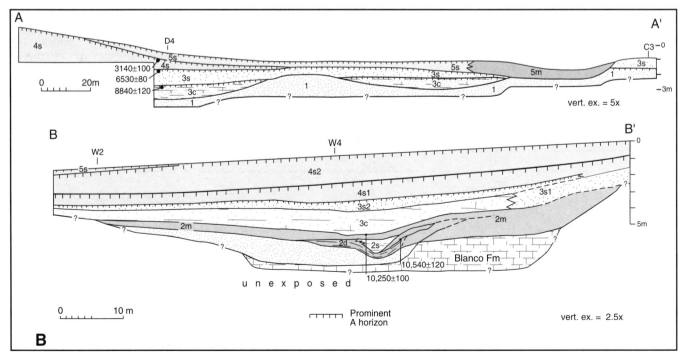

for recovery and analytical methods). These efforts met with mixed success.

The occurrence of vertebrate and invertebrate remains varies from site to site and from stratum to stratum, depending on the vagaries of original abundance and subsequent preserva-tion. Adequate recovery therefore is dependent on fortuitous encounters and, in the case of the microvertebrates, beetles, gastropods, and bivalves, also usually requires washing of large volumes of sediment (Johnson, 1987b; Elias and Johnson, 1988). During the present draw study no large vertebrate or microvertebrate remains were found, and recovery of beetles and pollen was very limited. Gastropod and bivalve remains are somewhat more abundant in the draws and the results of the phy-tolith and ostracode analyses are promising. Diatoms are the most abundant of fossils in the valley fills, occurring as beds of diatomite or in diatomaceous muds and other paludal and lacus-trine deposits. The results of the paleontological and paleo-botanical research follows.

Fossil biosilicates

by Steven Bozarth

The goals of this investigation were to determine the feasi-bility of recovering fossil biosilicates (opal phytoliths, siliceous algal bodies, and sponge spicules) from the valley fill and, if suc-cessful, to reconstruct the vegetative histories of the sample sites. The study included samples from eight sites in Running Water Draw, Yellowhouse Draw, Blackwater Draw, Mustang Draw, and Sulfur Draw.

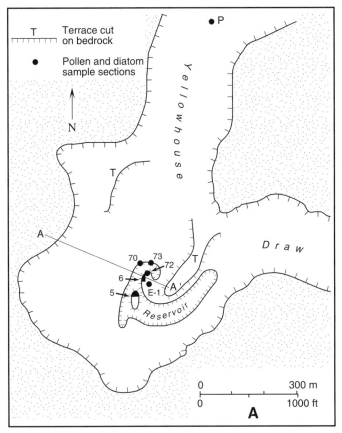

Figure 28. A, Lower Yellowhouse Draw at the Lubbock Lake site (Fig. 4) with the locations of the terraces cut on bedrock, the reservoir exca-vated in 1936, pollen, diatom, and beetle sampling sites (pollen from trenches 70 and 72, and locality P; diatoms from trenches 73 and E-1; insects from trench 73, and excavation areas 5 and 6), and location of section A–A' across the valley (Fig. 28B). See Figure 12A for key to symbols. B, Stratigraphic cross section A–A' across Yellowhouse Draw at the Lubbock Lake site (modified from Holliday, 1985c, fig. 3; Fig. 28A). See Figure 6 for key to symbols.

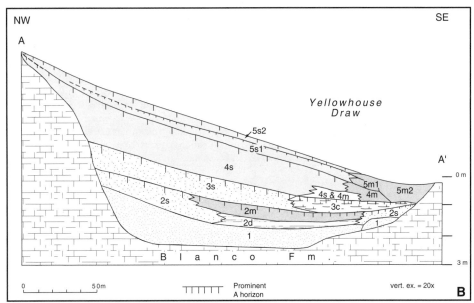

Opal phytoliths are silicified plant cells, cell walls, and intercellular spaces. Monocotyledons, particularly the Poaceae (grasses), produce a wide variety of morphologically distinct phytolith forms. One group of grass phytoliths, called short cells, are diagnostic at the subfamily level (Twiss, 1987). Nongrass monocots such as *Cyperus* (sedge) and *Scirpus* (bulrush) also produce diagnostic phytoliths. These recognizable phytoliths generally are well preserved in sediment. Dicotyledons (deciduous shrubs and trees, forbs, and weeds) produce diagnostic phytoliths as well (Rovner, 1971; Geis, 1973; Wilding et al., 1977; Bozarth, 1992). In contrast to the monocots, however, most dicot phytoliths are not usually preserved (Wilding and Drees, 1974; Bozarth, 1992). Several types of taxonomically useful phytoliths are produced in the *Pinaceae* (pine family; Norgren, 1973; Klein and Geis, 1978; Bozarth, 1988, 1994), and may be recovered through sample processing. Phytoliths can be isolated from sediment samples and analyzed to reconstruct paleoenvironments (Fredlund et al., 1985; Piperno, 1988; Bozarth, 1986).

Phytolith preservation generally was good in the samples from the Edmonson, Flagg, Lubbock Landfill (stratum 4s A-horizon), Mustang Springs, and Sundown sites. However, significant amounts of phytoliths were not present in the samples from the Brooks, Evans, Gibson, and Lubbock Landfill (4s) sites. The poor phytolith preservation in the latter samples may be due to high alkalinity, as the sediment samples had a pH of 8.4 or greater (Bozarth, 1989). Phytolith concentration varied considerably among the five sites with good preservation. Phytolith concentration was relatively low at Mustang Springs but relatively high in the Edmonson, Flagg, Lubbock Landfill (stratum 4s A-horizon), and Sundown sites.

Diatom and statospore (spherical cysts with siliceous walls produced in golden algae) preservation was generally good in all samples. Concentration of these siliceous algal bodies was relatively high in samples from strata 1s, 2d, 2m, and 4 (fill from an Archaic well; Meltzer, 1991) at Mustang Springs. Diatom concentration also was high in the samples from the Brooks and Gibson sites. Diatoms are treated in greater detail by Winsborough elsewhere in this section.

The relative frequency of all fossil biosilicates types is given in Table 14 for those samples with good phytolith preservation. Relative frequency data for fossil phytoliths alone are presented separately in Table 15 so that the presence of the other biosilicates (i.e., diatoms, statospores, and sponge spicules) does not complicate the interpretation of the phytoliths. Also separated were the relative frequency data concerning short-cell grass phytoliths in order to factor out those phytoliths formed in local hydrophytes (Table 16).

The samples from Mustang Springs (Tables 14–16) have two distinct biosilicate sections. The upper section (strata 4s, 5m, and 5s2) is dominated by grass phytoliths, the majority (85%, 89%, and 93%, respectively) of which are Chloridoids (a short-grass subfamily that flourishes in areas of warm temperature and low available soil moisture). These three phytolith assemblages were most likely produced by grama-buffalograss (*Bouteloua-Buchloe*) prairies as they are very similar to a modern analog from a short-grass prairie in western Kansas. The low frequency of opaque phytoliths, a well-preserved phytolith type common in deciduous trees and conifers, supports this hypothesis. These data indicate that the vegetation in the study area was a short

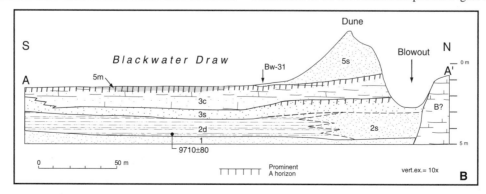

Figure 29. A, Middle Blackwater Draw in the area of the Gibson site (Fig. 4) with the locations of section A–A′ across the valley (Fig. 29B), all cores taken outside the blowout, and the surrounding Muleshoe Dunes. The blowout is probably the Marks Beach site of Honea (1980). See Figure 12A for key to symbols. B, Stratigraphic cross section A–A′ across middle Blackwater Draw at the Gibson site (Fig. 29A) showing the location of the radiocarbon sample. See Figure 6 for key to symbols.

grassland when the sediments represented by these samples were deposited. This hypothesis is supported by a low frequency of diatoms, which is characteristic of dry depositional conditions.

The lower section (strata 1, 2d, and 2m) is distinguished by relatively high frequencies of diatoms indicating wet or perhaps ponded conditions. The presence of sedge and/or bulrush phytoliths in each of these samples, and the presence of one sponge spicule in stratum 2m, provides supporting evidence of a wet or ponded depositional environment.

The lower section also is distinguished from the upper by the presence of polyhedral phytoliths and Type 1 and 2 phytoliths. Polyhedrons have been reported to be characteristic of conifers, but at Mustang Springs they were probably produced by herbaceous plants as evidenced by the absence or paucity of

opaque phytoliths indicative of deciduous and coniferous woodlands. The interpretation of the Type 1 and 2 phytoliths is problematic because they have not been reported in the literature. The Type 1 phytolith may be formed in a bulrush species as it has truncated cone-shaped projections similar to those of *Scirpus pallidus* phytoliths, but on rectangular bases instead of circular wavy bases. The Type 2 phytoliths may be from a hydrophilous plant species, as well, because they occur only in those samples with relatively high frequencies of diatoms. As in the upper section, the high frequencies of Chloridoid short cells in the lower section strongly suggests that the areal upland vegetation was short grassland.

The Archaic well-fill sample collected in stratum 4 of Mustang Springs also has a high frequency of diatoms indicating a wet depositional environment. The areal vegetation no doubt remained a short grassland when that stratum was deposited, based on the high frequency of Chloridoid phytoliths.

The samples from the Flagg, Sundown, Lubbock Landfill, and Edmonson sites all have very similar biosilicate assemblages (Tables 14–16), indicating that similar types of vegetation were growing in the separate areas during the periods of landscape stability associated with the surfaces of the Yellowhouse soil and the Lubbock Lake soil. As in the Mustang Springs samples, the high frequencies of Chloridoid short cells strongly indicate that shortgrass prairie was the dominant type of vegetation. This hypothesis is supported by the absence or paucity of opaque phytoliths common in arboreal species. The extremely low frequency of

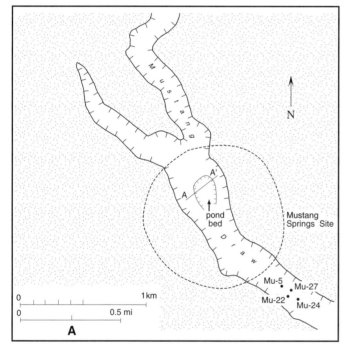

Figure 30. A, Lower Mustang Draw in the area of the Mustang Springs site (encompassed by dotted line; Fig. 4) with the locations of section A–A′ across the draw (Fig. 30B), the headcut defining the historic pond bed, and locations of cores outside of the site area proper. See Figure 12A for key to symbols. B, Generalized stratigraphic cross section A–A′ across Mustang Draw at the Mustang Springs site (Fig. 30A) showing locations of radiocarbon samples and schematically illustrating the location of hand-dug, prehistoric wells (based on Meltzer, 1991, fig. 3; and data provided by D. J. Meltzer, personal communication, 1992). See Figure 6 for key to symbols.

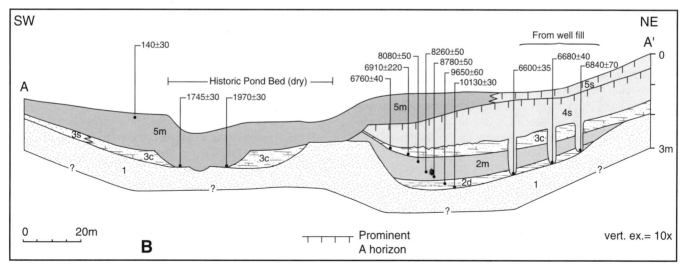

TABLE 14. PERCENTAGE OF BIOSILICATES FROM VALLEY FILL IN THE SOUTHERN HIGH PLAINS

Sample	14C Ages (yr B.P.)	Chloridoideae[1]	Aristida[2]	Panicoideae[3]	Pooideae[4]	Stipa[5]	Cyperus[6]	Scirpus[7]	Asteraceae[8]	Polyhedrons	Spiney spheroids	Opaque phytoliths	Type 1 phytoliths[9]	Type 2 phytoliths[10]	UI[11] phytoliths	Diatoms	Statospores	Sponge spicules	Specimens counted
Mustang Springs																			
Stratum 5s2		77[12]	2	2	1	-[13]	-	-	-	-	-	-	-	-	7	1	9	-	82
Stratum 5m	2,000	79	1	-	7	1	-	-	-	-	-	1	-	-	5	2	3	-	74
Stratum 4s		78	2	4	7	1	-	-	-	-	-	3	-	1	3	2	-	-	105
Stratum 4 (well-fill)	6,700	30	-	-	3	-	-	-	-	1	-	1	-	-	1	50	13	-	145
Stratum 2m		14	1	2	2	1	1	-	-	3	-	-	1	+[14]	1	68	6	+	391
Stratum 2dG	8,300	30	+	1	2	+	1	+	-	3	-	-	2	+	2	43	13	-	359
Stratum 2dB		19	1	1	2	+	-	1	-	11	-	-	2	1	3	49	10	-	198
Stratum 2dA	10,000	16	1	1	1	1	2	-	+	9	-	1	7	5	3	40	12	-	146
Stratum 1s		42	1	1	4	-	-	1	-	9	-	4	6	6	11	7	7	-	81
Flagg Site																			
Stratum 4, A horizon	770	89[12]	1	2	4	-[13]	-	-	-	-	-	-	-	-	3	-	+[14]	-	211
Sundown Site																			
Stratum 4, A horizon[15]	2,100	90	1	2	2	+	-	-	-	-	-	+	-	-	2	1	1	-	211
Lubbock Landfill																			
Stratum 4, A horizon	3,100	88	2	2	3	+	-	-	-	-	-	+	-	-	3	-	+	-	209
Edmonson Site																			
Stratum 3c, A horizon	8,000	81	1	3	4	+	-	-	-	2	+	+	-	-	2	3	2	-	240

[1] Short-grass subfamily; [2] genus in short-grass subfamily; [3] tall-grass subfamily; [4] cool-moist season grass subfamily; [5] genus in cool-moist season subfamily; [6] sedge; [7] bulrush; [8] sunflower family; [9] epidermal cells with truncated cone projections; [10] globular bodies with short blunt processes; [11] unidentified; [12] relative frequency rounded to nearest whole number; [13] "-" no observations; [14] "+" present at less than 0.5 percent; [15] lower half of cumulic A horizon.

diatoms and statospores and lack of sponge spicules in all four samples indicates dry depositional conditions.

Diatom and statospore preservation generally was good at all study sites. Relatively high frequencies of these two types of siliceous algal bodies and the presence of sedge and bulrush phytoliths in the lower section (strata 1s–2m) of Mustang Springs indicate that these strata were formed under locally wet or ponded conditions. High frequencies of diatoms and statospores were also found in the Gibson and Brooks sites, indicating a similar depositional environment.

Phytolith preservation was adequate for analysis at the Mustang Springs, Flagg, Sundown, Lubbock Landfill, and Edmonson sites. Relatively high frequencies of Chloridoid grass phytoliths strongly indicate that short grassland was the dominant areal vegetation type at Mustang Springs from ca. 10,000 yr B.P. to the present, at the Edmonson site ca. 8,000 yr B.P., at Lubbock Landfill ca. 3,100 yr B.P., at the Sundown site ca. 2,100 yr B.P., and at the Flagg site ca. 770 yr B.P.

Collectively, these data suggest that short grasslands covered most of the Southern High Plains during the Holocene. Whereas the draws concurrently afforded pond or marsh environments supporting aquatic vegetation.

Insect fossils

by Scott A. Elias

Quaternary insect fossils have become an important tool in the reconstruction of the timing and intensity of environmental change. Fossil insects have proven to be sensitive, reliable proxy data for the reconstruction of past environments (Coope, 1970; Elias, 1994). A pilot study at Lubbock Lake indicated that, under proper sedimentological, geochemical, and recovery conditions, fossil insects could be secured from late Quaternary valley fill on the Southern High Plains (Elias and Johnson, 1988). During the 1988–1992 study of the draws small assemblages of insect fossils were identified in organic deposits at Lubbock Lake and Plainview. Many of these deposits were paleosols, and insect preservation in these sediments was poor.

Fossil beetles identified from the Lubbock Lake site (Table

TABLE 15. PERCENTAGE OF PHYTOLITHS FROM VALLEY FILL IN THE SOUTHERN HIGH PLAINS

Sample	14C Ages (yr B.P.)	Chloridoideae[1]	Aristida[2]	Panicoideae[3]	Pooideae[4]	Stipa[5]	Cyperus[6]	Scirpus[7]	Asteraceae[8]	Polyhedrons	Spiney spheroids	Opaque phytoliths	Type 1 phytoliths[9]	Type 2 phytoliths[10]	UI[11] phytoliths	Total phytoliths
Mustang Springs																
Stratum 5s2		85[12]	3	3	1	-[13]	-	-	-	-	-	-	-	-	8	74
Stratum 5m	2,000	83	1	-	7	1	-	-	-	-	-	1	-	-	6	70
Stratum 4s		80	2	4	7	1	-	-	-	-	-	3	1	-	3	103
Stratum 4 (well-fill)	6,700	81	-	-	9	-	-	-	-	2	-	3	-	-	3	54
Stratum 2m		56	4	6	7	3	3	-	-	12	-	-	4	1	4	100
Stratum 2dG	8,300	70	1	3	5	1	2	1	-	8	-	-	4	1	4	156
Stratum 2dB		47	2	2	3	1	-	2	-	27	-	-	5	2	6	81
Stratum 2dA	10,000	35	3	3	3	3	4	-	1	14	-	1	14	12	7	69
Stratum 1s		49	1	1	4	-	-	1	-	10	-	4	7	7	13	69
Flagg Site																
Stratum 4, A horizon	770	90[12]	1	2	4	-[13]	-	-	-	-	-	-	-	-	3	210
Sundown Site																
Stratum 4, A horizon[15]	2,100	92	1	2	2	+	-	-	-	-	-	+[15]	-	-	2	207
Lubbock Landfill																
Stratum 4, A horizon	3,100	88	2	2	3	+	-	-	-	-	-	+	-	-	3	208
Edmonson Site																
Stratum 3c, A horizon	8,000	84	1	4	4	+	-	-	4	-	+	+	-	-	2	227

1 Short-grass subfamily; 2 genus in short-grass subfamily; 3 tall-grass subfamily; 4 cool-moist season grass subfamily; 5 genus in cool-moist season subfamily; 6 sedge; 7 bulrush; 8 sunflower family; 9 epidermal cells with truncated cone projections; 10 globular bodies with short blunt processes; 11 unidentified; 12 relative frequency rounded to nearest whole number; 13 "-" no observations; 14 lower half of cumulic A horizon; 15."+" present at less than 0.5 percent.

17) include ground beetles (Carabidae), dung beetles (Scarabaeidae), and weevils (Curculionidae). Ground beetles are mostly predators, feeding on insects and other arthropods. *Calosoma porosifrons* is a caterpillar hunter. This beetle was found in stratum 2d, dated approximately 10,200 yr B.P. (Elias and Johnson, 1988). Its known modern range is restricted to the state of Durango, Mexico, in sites ranging in elevation from valley floor to mountains, where the vegetation grades from desert grassland to pine-oak forest (Gidaspow, 1959). Annual precipitation in that region ranges from 400–600 mm and mean July temperatures are approximately 25° C (Tamayo, 1962). Ground beetles of the genus *Amara* occur in stratum 2d (11,000–10,000 yr B.P.), but are not environmentally diagnostic.

The ground beetle *Agonum* of the subgenus *Rhadine* was found in stratum 2s, below a level dating to 10,000 yr B.P. The sand in 2s at this locality, however, was derived from a nearby Pleistocene terrace and contains redeposited Pleistocene fauna. The age of the beetle remains is therefore uncertain. Most species of this subgenus are cave dwellers that prey on cave crickets and other cave invertebrates. Its presence at Lubbock Lake is some-

what enigmatic, because there are no known caves or natural rock shelters in the vicinity of the site.

The dung beetles found at Lubbock Lake, both from stratum 2d, could not be specifically identified. The genus *Aphodius* is very widespread and feeds on many kinds of dung. The genus *Aegialia* is likewise ubiquitous, but these beetles feed on detritus. *Aegialia* are found on sand dunes and other sandy substrates (Gordon and Cartwright, 1988).

The weevils from Lubbock Lake include three genera. A specimen of the genus *Listronotus* was found in 10,200-yr-old sediments of 2d. These beetles feed on various subaquatic plants, including sedges (*Carex* spp.), arrowhead (*Sagittaria* spp.), and knotweed (*Polygonum* spp.; Kissinger, 1964). The weevil, *Anametis subfusca*, was identified from a layer in stratum 2d (11,000–10,000 yr B.P.). This beetle lives today in oak-pinyon-juniper woodlands in Colorado, New Mexico, Texas, and Arizona (C. O'Brien, written communication, 1990), where it feeds on a variety of plants, including juniper (*Juniperus* spp.), snakeweed (*Xanthocephalum* sp.), and cliffrose (*Cowania ericaefolia*; R. Anderson, Canadian National Museum of Natural Sciences,

TABLE 16. RELATIVE FREQUENCY OF GRASS PHYTOLITHS FROM VALLEY FILL IN THE SOUTHERN HIGH PLAINS

Sample	¹⁴C Ages (yr B.P.)	Chloridoideae[1]	Aristida[2]	Panicoideae[3]	Pooideae[4]	Stipa[5]	Total phytoliths
Mustang Springs							
Stratum 5s2		93	3	3	2	-[6]	68
Stratum 5m	2,000	89	2	-	8	2	65
Stratum 4s		85	2	4	7	1	96
Stratum 4 (well-fill)	6,700	90	-	-	10	-	49
Stratum 2m		74	5	8	9	4	76
Stratum 2dG	8,300	87	2	4	6	2	126
Stratum 2dB		83	4	4	7	2	46
Stratum 2dA	10,000	75	6	6	6	6	32
Stratum 1s		87	3	3	8	-	39
Flagg Site							
Stratum 4, A horizon	770	92	1	2	4	-	204
Sundown Site							
Stratum 4, A horizon[7]	2,100	95	1	2	2	+[8]	201
Lubbock Landfill							
Stratum 4, A horizon	3,100	92	2	2	3	+	201
Edmonson Site							
Stratum 3c, A horizon	8,000	91	1	4	5	1	212

[1] Short-grass subfamily; [2] genus in short-grass subfamily; [3] tall-grass subfamily; [4] cool-moist season grass subfamily; [5] genus in cool-moist season subfamily; [6] "-" no observations; [7] lower half of cumulic A horizon; [8] "+" present at less than 0.5 percent.

TABLE 17. FOSSIL INSECT LIST, LUBBOCK LAKE SITE,* IN MINIMUM NUMBER OF INDIVIDUALS PER SAMPLE

Taxon	Stratum				
	2d[†]	2d[§]	2s**	2m[‡]	"X"
COLEOPTERA					
Carabidae					
Calosoma porosifrons Bates		1			
Amara sp.	3				
Agonum (Rhadine) sp.			1		
Genus et sp. indet.				1	1
Scarabaeidae					
Aphodius sp.		2			
Aegialia sp.		2			
Curculionidae					
Listronotus sp.		1			
Ophryastes sp.					1
Anametis subfusca Fall	1				
Genus et sp. indet.		2			

*Figures 4, 28.
[†]11,000-10,000 yr B.P., Trench 73 (Fig. 28A).
[§]ca. 10,200 yr B.P., excavation area 6 (Fig. 28A).
**Sand redeposited from Pleistocene terrace.
[‡]10,000-9,000 yr B.P., excavation area 5 (Fig. 28A).
"x" is informal stratigraphic designation for eolian deposit <1,000 yr B.P. on strath terrace above the valley fill (Fig. 28A).

Ottawa, written communication, July 1990). Finally, a specimen of *Ophryastes* was identified from an eolian deposit (informally designated stratum X) that blankets the strath terrace above the valley fill, dating to within the last 1,000 years. This genus is mostly found in arid regions, where it feeds on such plants as sagebrush (*Artemisia* spp.), saltbrush (*Atriplex* spp.), tarbush (*Flourensia cernua*), and creosote bush (*Larrea tridentata*; Kissinger, 1964).

The beetle fossils, while few, provide some evidence of past environments at the Lubbock Lake site. At about 10,000 yr B.P., conditions at the site were markedly cooler and wetter than present, allowing the growth of shrubs and other plants found today only at higher elevations in the American Southwest. By about 1,000 yr B.P. and onwards, regional climate was essentially modern.

Holocene buried soils from the Plainview site also yielded a small fossil beetle fauna (Table 18). Nearly all of the beetle species were identified from soils less than 2,000 years old. The middle of stratum 5s2 yielded two genera of darkling beetles (Tenebrionidae), indicative of dry, upland environments. Also in this layer was the seed beetle (Bruchidae), *Acanthoscelides*. This genus feeds on a variety of legumes, including mesquite

(*Prosopis* spp.). An indeterminate genus of Curculionidae was found in stratum 5s2, as well.

The A horizon in 5s2, dated at less than 1,000 yr B.P., yielded several species of beetles. The ground beetle, *Cratacanthus dubius*, is widely distributed across the United States, as far west as Colorado (Elias, 1987). *C. dubius* lives in open, dry country, often in cultivated fields (Lindroth, 1968). Also contained in the buried soil horizon were several taxa of dung beetles. One specimen of *Aphodius* was recovered that closely resembles *A. crenicollis*. This species lives today in the western United States, and is frequently found in rodent burrows, including those of ground squirrels (*Citellus* and *Spermophilus*) and prairie dogs (*Cynomys* spp.). Another dung beetle found in this buried soil was *Onthophagus pennsylvanicus*. This species is widespread in the eastern and central United States, as far west as Texas and Colorado. It feeds on many kinds of dung (Howden and Cartwright, 1963).

Finally, the weevil, *Smicronyx scapalis*, is found today in southern Kansas, Oklahoma, and Texas (Anderson, 1962). Its principal host plant is the daisy, *Haplopappus rubignosus*, but it has also been collected from gumweed (*Grindelia squarrosa*).

While the fossil beetle evidence from Plainview is scanty at best, it suggests that conditions in the last 2,000 years have been similar to modern, with semiarid climate, dry soils, and vegetation including mesquite, grasses, and herbs. The dung beetle evidence indicates local populations of ground squirrels and prairie dogs, which were ubiquitous on the Southern High Plains, up to the twentieth century.

TABLE 18. FOSSIL INSECT LIST, PLAINVIEW SITE, TEXAS,* IN MINIMUM NUMBER OF INDIVIDUALS PER SAMPLE

Taxon	2m†	5s2 Middle§	5s2 Top**
COLEOPTERA			
Carabidae			
Amara sp.			1
Crataeanthus dubius Beauv.			1
Scarabaeidae			
Aphodius nr. *crenicollis* Fall			1
Aphodius sp.			1
Onthophaqus pennsylvanicus Har.			1
Genus et sp. indet.			1
Histeridae			
Genus et sp. indet.			1
Tenebrionidae			
Coniontis sp.		1	1
Genus et sp. indet.		1	1
Bruchidae			
Acanthoscelides sp		1	
Curculionidae			
Smicronyx scapalis (LeC.)			1
Genus et sp. indet.	1	1	

*Figures 4, 26.
†ca. 8,900 yr B.P.
§ca. 2,000-1,000 yr B.P.
**<1,000 yr B.P.

Pollen

by Stephen A. Hall

This report is a summary of attempts to recover pollen from the late Quaternary valley fill of the Southern High Plains (Table 19). Although standard and nonstandard techniques were applied (GSA Data Repository 9541, Appendix C), the results are discouraging.

Pollen samples were collected at the Clovis site from stratum 2 (undifferentiated) in earlier studies by Howard (1935a). Paul Sears analyzed the material and recovered only a few pollen grains that were unsuitable for interpretation (Hester, 1972, p. 24). Hafsten (1961) studied samples from strata 1 and 2d and likewise found little pollen. Subsequently, Schoenwetter (1975) reported some pollen frequencies based on low to moderate counts from two stratigraphic sections, although 17 of 36 samples were barren of pollen.

A minor controversy surrounds palynological research at the Lubbock Lake site, discussed by Bryant and Schoenwetter (1987), where a number of analysts recovered pollen assemblages marked by low concentration and poor preservation (e.g., Hafsten, 1961, p. 66, 75; Shoenwetter, 1975, p. 112),

TABLE 19. HIGH PLAINS DRAW LOCALITIES INSPECTED FOR POLLEN*

Locality†	Comments on Palynology§
BLACKWATER DRAW	
Bailey Draw, Bw-39	
Stratum 3c, 1 sample (138-143 cm)	Few charred particles, some fungi, no pollen.
Stratum 1, 1 sample (229-264 cm)	Abundant, diverse fungal bodies, almost no charred particles, no pollen.
Clovis site, South Bank	
Stratum 2s, 1 sample	Abundant charred particles and wood charcoal, no pollen.
Davis site, Bw-45	
Stratum 2m, 1 sample (300-330 cm)	Abundant charred particles with charcoal, few pollen grains (*Pinus, Picea*) broken and corroded, low concentration.
Progress Draw, Bw-42	
Stratum 2m, 1 sample (240-285 cm)	Numerous charred particles, some fungi, no pollen.
Tolk site, Bw-36	
Stratum 2m, 1 sample (152-200 cm)	Moderate number of charred particles, no pollen.
MUSTANG DRAW	
Glendenning site, Mu-3	
Stratum 4s, 1 sample (0-10 cm)	Moderately abundant pollen grains, all poorly preserved; numerous charred particles.
Stratum 3c (A horizon) 1 sample (170 cm)	Some charred particles, fungi, no pollen.
Stratum 3c, 1 sample (190 cm)	Some charred particles, no pollen.
Stratum 1, 1 sample (330 cm)	No charred particles, no pollen.
Mustang Springs, Mu-1	
Stratum 2d, 1 sample (325-337 cm)	Numerous charred particles, no pollen.
RUNNING WATER DRAW	
Edmonson site, Rw-19	
Stratum 3c, 1 sample (350-360 cm)	Small charred particles, some fungi, no pollen.
Strata 3c, 1 sample (360-368 cm); 2m, 1 sample (369-400 cm)	A few small charred particles, a few fungi, some pollen (5 *Pinus,* 1 *Picea*), very low pollen concentration.
Flagg site, Rw-11	
Strata 4s, 1 sample (255-261); 3c, 1 sample (261-270); 2d, 1 sample (270-283); 2m, 1 sample (283-300)	Numerous large charred particles, some fungi, two broken *Pinus* grains.
Plainview site, Pit 3	
Profile 9, stratum 2m 1 sample	Some charcoal fragments, some fungi, no pollen.
Profile 12, stratum 3m 1 sample	Some charred particles, no pollen.
Profile 12, stratum 2m 1 sample	Some small charred particles, some fungi, no pollen.

*All samples collected by V. T. Holliday.
†Figures 3, 4.
§All residues contain numerous lycopod spores introduced as a spike in each sample during laboratory processing.

currently regarded as the mark of material of questionable reliability for vegetation reconstruction. The writer examined two samples from exposures of strata 2d and 2m (Table 20; Fig. 28A) and found a few pollen grains, all of which were corroded and with low concentrations. One of the samples had 100% *Pinus* and *Picea* pollen and is interpreted as a product of differential preservation (Table 20). Because of the possibility that pollen may become destroyed upon weathering of exposed outcrops, diatomite from two cores were analyzed by the writer for pollen (Table 20; Fig. 28A). Unfortunately, the core material yielded only a small amount of poorly preserved pollen grains, similar to the corroded pollen assemblages from outcrops.

Samples from the Bailey Draw, Progress Draw, Clovis, Davis, Tolk, Edmonson, Flagg, and Glendenning sites were inspected for pollen. Virtually all of the sediment was found barren of pollen (Table 19). An exception to the general pattern of pollen destruction was discovered at the Plainview site where a thin zone of organic-rich clay in stratum 2m (dated ca. 8,900 yr B.P.) contains a low abundance of moderately well preserved pollen (Table 21). The low counts from the single-sample

assemblage indicate a grassland vegetation without *Artemisia* but with a moderate amount of pinyon-pine pollen (22%) that could represent the presence of small populations of pinyon pine along the High Plains escarpment during the early Holocene.

Pollen assemblages characterized by low pollen concentration, low taxa diversity, and high proportions of corroded grains, may have been altered by differential destruction of pollen and, as a result, may not be reliable for vegetation reconstruction (Bryant and Hall, 1993). Most of the late Quaternary sediments that fill the draws of the Southern High Plains unfortunately fall in this situation.

Stable isotopes

by Vance T. Holliday

Stable-carbon isotopes from organic-rich sediment, from soil carbonate, and from bone allow inferences to be made regarding past vegetation, and therefore the environment, because there is a strong and positive correlation between δ^{13}C

TABLE 20. POLLEN DATA FROM LUBBOCK LAKE TRENCHES AND CORES*

Pollen Taxa	Pollen Counts	Percentages	Pollen Counts	Percentages
		TRENCHES[†]		
	Trench 70		Trench 72	
Pinus	1	25	43	93.5
Picea	1	25	3	6.5
Chenopodiineae	1	25	-	-
Indeterminable	1	25	-	-
Pollen sum	4		46	
Spike count	125		110	
Pollen concentration (grains/gram)	37		427	
Corroded and/or degraded grains	4	100	46	100
Botryococcus (algae)	-		1	
		CORES[§]		
	Sample A		Sample B	
Pinus	26	63.4	26	68.4
Picea	4	9.8	1	2.6
cf. *Juniperus*	4	9.8	-	
cf. *Quercus*	1	2.4	-	
Poaceae	2	4.9	-	
Typha	-		5	13.2
Indeterminable	3	7.3	6	15.8
Unknown	1	2.4	-	
Pollen sum	41		38	
Spike count	121		143	
Pollen concentration (grains/gram)	333		403	
Corroded and/or degraded	29	70.7	22	57.9
Botryococcus (algae)	-		2	

*Figures 4, 28A.
[†]Stratum 2d; collected by S. A. Hall and V. T. Holliday from trenches along northwest margin of reservoir (Fig. 28A).
[§]Stratum 2d; collected by V. T. Holliday from cores 0.5 km above reservoir (locality P; Fig. 28A)

TABLE 21. POLLEN DATA FROM THE PLAINVIEW SITE, PROFILE 12*

Pollen Taxa	Pollen Count	Percentages
Pinus sp.	4	
Cf. *Pinus edulis*	22	
Pinus sp., not cf. *P. edulis*	2	
Pinus, total	28	28
Cf. *Populus*	3	3
Poaceae	5	5
Ambrosia	19	19
Liguliflorae	10	10
Other Asteraceae	12	12
Chenopodiineae	6	6
Cf. Brassicaceae	5	5
Onagraceae	1	1
Cyperaceae	2	2
Indeterminable	6	6
Unknown	3	3
Pollen sum	100	
Spike count	140	
Pollen concentration (grains/gram)	497	
Total corroded/degraded grains, all taxa	42	42

*Collected by V. T. Holliday; Stratum 2m, ca. 8,900 yr B.P., (Figs, 4, 26).

and the proportions of C_3 and C_4 biomass (Cerling, 1984; Bowen, 1991). Given the well-dated and essentially continuous record of sedimentation and soil formation preserved in the draws of the Southern High Plains, a pilot study was conducted to establish a record of C-isotopic trends through time as an indicator of environmental change. The resulting data complement that available from the sedimentological, pedological, paleontological, and paleobotanical studies.

Isotopic data are available from organic-rich sediments and soils at four sites (Table 22). The study focused on organic carbon (OC) because OC in sediments and soils is the most ubiquitous form of carbon in the valley fill and because OC representing much of the late Quaternary was found in similar settings (valley axes) and probably formed in similar environments (moist, well-vegetated lowlands) at several widely scattered sites. OC isotopic data are available from previous studies at Lubbock Lake (Haas et al., 1986, Table 4; E. Johnson, unpublished data) and Mustang Springs (Meltzer, 1991, table 1). Additional OC isotopic data also were gathered from the late Quaternary sections at Plainview and the Lubbock Landfill. Stafford (1984) conducted a study of $\delta^{13}C$ values in collagen of *Bison* bone from Lubbock Lake. Inorganic carbon from calcium carbonate proved unsatisfactory for study because samples from similar environmental settings and spanning the Holocene were not found. Most early Holocene carbonate was in paludal sediment and most middle and late Holocene carbonate was pedogenic and from well-drained soils.

Some trends are apparent in a comparison of the C-isotopic

data (Fig. 31), but several aspects of the isotopic characteristics of the samples must first be considered before interpretations are made. The $\delta^{13}C$ values for OC and from associated bone will vary because the carbon fractionates differently in plants and in animals. Organic matter from C_3 plants (cool season grasses, most aquatic plants, and all trees) has a mean $\delta^{13}C$ of approximately –26 per mil and organic matter from C_4 (mainly warm season grasses) averages –13 per mil (Bowen, 1991, p. 129). C_4 also shows more favorable physiological performance under warm and dry conditions than do C_3 plants. The $\delta^{13}C$ values in the collagen of *Bison* bones generally are similar to the values found in plant material: bones of individuals grazing on plants dominated by C_3 species contain about –22 per mil, while diets dominated by C_4 plants will produce values of –13 per mil (Tieszen, 1994). Moreover, Stafford (1984, p. 113–114) concluded that the $\delta^{13}C$ values from *Bison* bone "will reflect an average of the carbon ingested during at least the last 5–10 years of the animal's life. Seasonal fluctuations in grassland composition or an animal's migration into different grasslands should not alter the long-term carbon composition in bone collagen."

Plant types and associated $\delta^{13}C$ values also can vary within the draws. The stratigraphic and pedologic data discussed previously show that at any one time the local environment could vary from the axis to the margin of the draws. The valley axes historically and prehistorically often were wetter, more poorly drained, and higher in biomass production than were valley margins (as indicated by the organic-matter content and color of valley-axis A-horizons and sediments). Such settings favor growth of C_3 plants such as trees, most shrubs, and temperate grasses. The valley margins, in contrast, are and were much drier, and likely favored growth of the more arid adapted C_4 grasses. In addition, on long-stable land surfaces (denoted by buried A-horizons) there were undoubtedly variations in plant types at any one location over time due to changes in microenvironment such as soil moisture. If proportions of C_4 to C_3 plants changed during accumulation of OC, then the resulting $\delta^{13}C$ values will reflect a mixed signal of C_4 biomass. In the buried A-horizons, as with the radiocarbon dating of their humates, most of the organic carbon remaining will likely be from the time immediately predating burial of the surface due to oxidation and turnover of older carbon.

The $\delta^{13}C$ values for the latest Pleistocene (< 12,000 yr B.P.) and earliest Holocene (> 8,000 yr B.P.) samples of OC and bone (Fig. 31) are relatively light, that is, the data suggest relatively high amounts of C_3 biomass. Most of the isotopic ratios > 8,000 yr B.P. are lighter than the values 8,000–500 yr B.P., though there are fewer data points for the middle Holocene and early late Holocene samples. The change from lighter to heavier isotopic composition occurs between 10,000 and 7,500 yr B.P. At Plainview, the shift in OC $\delta^{13}C$ appears to happen gradually before 8,800 yr B.P., though the data are extremely sparse. The two sites providing the most data, Lubbock Lake and Mustang Springs, document a very abrupt shift in $\delta^{13}C$ of organic material: ca. 8,200 yr B.P. at Lubbock Lake and slightly less than 7,000 yr B.P. at Mustang Springs. The shift to heavier $\delta^{13}C$ ratios in OC at Lubbock Landfill, based on few data points, occurs around 8,500 yr B.P., a little sooner than at Lubbock Lake or Mustang Springs. Very generally, grasses and possibly other plants indicative of cooler (and perhaps more moist) conditions dominated the draws in the early

Holocene, but were replaced by grasses suggestive of warmer (and possibly direr) conditions in the middle Holocene.

The change to heavier $\delta^{13}C$ ratios in *Bison* bone occurs sometime between 8,500 and 5,000 yr B.P. (Table 22A; Fig. 31); no isotopic data are available on bone for the intervening years. At 5,200–5,000 yr B.P., however, there is considerable range in the $\delta^{13}C$ values of the bone. This variability is probably due to the effects of organic content, degree of weathering, and inorganic contamination (Stafford, 1984, p. 119).

OC-isotopic values for most late Holocene samples are indicative of C_3 grasses, although there are relatively few data points except at Lubbock Lake (Table 22A). The majority of these Lubbock Lake samples are < 2,000 years old because there was little late Holocene sedimentation. All of the $\delta^{13}C$ values for the period 2,000–500 yr B.P. are heavy, very similar to those of

TABLE 22A. $\delta^{13}C$ VALUES FOR ORGANIC-RICH SEDIMENTS AND SOILS AND FOR *BISON* BONE COLLAGEN AT THE LUBBOCK LAKE SITE*

Sample	Age (yr B.P.)	Stratum	$\delta^{13}C$
Sediments and Soils			
SMU-831[†]	390 ± 50	5m2	-25.5
SMU-2657	460 ± 60	5s	-28.0
SMU-2445	700 ± 50	5s	-14.9
SMU-2638	730 ± 40	5s	-16.3
SMU-2352	960 ± 55	5s	-15.3
SMU-2658	1,040 ± 50	5s	-17.3
SMU-1090[†]	1,270 ± 40	4s[§]	-14.0
SMU-2350	1,350 ± 40	5s	-15.6
SMU-1177[†]	1,550 ± 50	4s[§]	-15.5
SMU-2351	1,570 ± 70	5s	-15.2
SMU-2639	1,590 ± 40	5s	-16.0
SMU-2353	1,590 ± 90	5s	-15.0
SMU-1191[†]	2,070 ± 130	4s[§]	-15.7
SMU-1200[†]	5,270 ± 150	5m	-17.3
SMU-1093[†]	5,220 ± 50	3c	-14.0
SMU-1673	6,210 ± 70	4m	-14.5
SMU-1674	6,580 ± 60	3c	-14.2
SMU-1595	7,840 ± 70	2s	-14.8
SMU-830[†]	8,220 ± 240	2m (upper)	-22.8
SMU-1594	8,400 ± 70	2s	-18.8
SMU-1192	8,730 ± 240	2s	-18.0
SMU-1304	8,935 ± 90	2s (mid)	-23.6
SMU-829[†]	9,170 ± 80	2m (lower)	-22.7
SMU-1394	9,350 ± 60	2m (lower)	-23.3
SMU-1261[†]	9,950 ± 120	2s (lower)	-23.2
Bone**			
78LLP-6	500	5m	-7.9
78LLP-32	500	5m	-8.1
78LLP-2	700	5s1	-9.7
78LLP-22	5,000	4s (lower)	-11.4
78LLP-27	5,000	4s (lower)	-3.0
78LLP-17	5,000	4s (lower)	-14.8
78LLP-28	5,200	4m	-4.0
78LLP-8	5,200	4m	-10.2
78LLP-5	5,200	4m	-11.9
78LLP-25	8,500	2m (upper)	-15.8
78LLP-23	9,000	2m (mid)	-17.2
78LLP-29	9,800	2m (lower)	-14.3
78LLP-4	10,700	2d (mid)	-13.6
78LLP-10	11,200	1	-16.2

TABLE 22B. $\delta^{13}C$ VALUES FOR ORGANIC-RICH SEDIMENTS AND SOILS AT THE PLAINVIEW, LUBBOCK LANDFILL, AND MUSTANG SPRINGS SITES*

Sample	Age (yr B.P.)	Stratum	$\delta^{13}C$
Plainview			
SMU-1239	2,070 ± 130	4s	-15.5
SMU-2534	2,600 ± 40	5m	-15.5[†]
SMU-1234	3,880 ± 60	4s	-16.0
SMU-1349	6,770 ± 50	3s	-16.1
SMU-2341	8,860 ± 110	2m	-16.4
SMU-1359	10,940 ± 70	1m	-17.9
SMU-1367	11,970 ± 140	1m	-19.2
Lubbock Landfill[§]			
Beta-43007	3,140 ± 100	4s	-15.6
Beta-43008	6,530 ± 80	3si	-15.5
SMU-1675	8,130 ± 70	3c	-15.7**
Beta-43009	8,840 ± 120	3c	-19.4
Beta-57226	10,250 ± 100	2m	-19.6
Beta-61962	10,540 ± 120	2m	-17.6
Mustang Springs[‡]			
SMU-1589	140 ± 30	5m	-13.7
SMU-1785	1,745 ± 30	5m	-17.0
SMU-1588	1,970 ± 30	5m	-14.8
SMU-1971	6,600 ± 35	4s	-16.3
SMU-1800	6,680 ± 40	4s	-16.1
SMU-1786	6,840 ± 70	4s	-16.8
SMU-1783	6,760 ± 40	3c	-20.4
SMU-2173	6,915 ± 220	3c	-18.3
SMU-1784	8,080 ± 50	2m	-22.2
SMU-1587	8,260 ± 50	2m	-22.1
SMU-1799	8,780 ± 50	2m	-21.1
SMU-1586	9,650 ± 60	2d	-22.1
SMU-1585	10,130 ± 30	2d	-21.8

*Figures 4, 28.
[†]From Haas et al., 1986, Table 4.
[§]Sample from buried A horizon at top of stratum 4s.
**Modified from Stafford, 1984, Table 2; some dating revised from Stafford, 1984, based on additional C^{14} ages (Holliday et al., 1983, 1985).

*Figures 4, 26, 27, 30.
[†]Sample from Quincy Street locality, 1.3 km above dated section at Plainview (Fig. 26A).
[§]Samples determined by Beta are from Brown et al., 1993.
**Sample from BFI site, 2 km below dated sections at the Landfill.
[‡]From Meltzer, 1991, Table 1.

the middle Holocene. There is a dramatic shift toward lighter δ13C values for samples of OC dated to within the past 500 years.

The carbon-isotope values suggest several shifts in the vegetation composition of the Southern High Plains regionally and in the draws during the Holocene. Each per mil change represents a shift of about 7% in the ratio of C_3 to C_4 (Pendall and Amundson, 1990; Nordt et al., 1994). At Lubbock Lake and Mustang Springs, therefore, early Holocene vegetation probably was about 28% C_4 plants (mean δ13C value of about -22 per mil) and middle Holocene vegetation about 79% C_4 (mean δ13C value of about -15 per mil). Stafford (1984, p. 116) speculated that if the relationship between bone δ13C and grassland percentage C_4 is linear then C_4 grasses may have made up 45% of the grasslands at 12,500 yr B.P. and 65% at 8,000 yr B.P., compared to 95% C_4 grasses today. He notes, however, that a nonlinear relationship between bone and grassland composition results if *Bison* preferentially ingest more C_3 grasses. Data on such behavioral fractionation are not available.

The shift in early Holocene vegetation composition almost certainly is related to environmental change, but sorting out the direct effects of climate-to-vegetation versus climate-to-local-environment-to-vegetation is difficult. The sharp change to heavier isotopic signatures early in the Holocene at Lubbock Lake (ca. 8,200 yr B.P.) and Mustang Springs (ca. 6,800 yr B.P.) coincides with the geologic change from the stratum 2m depositional environment (low or no carbonate) to the stratum 3c depositional environment (moderate to high carbonate) deposition. Such a shift could and likely did produce vegetation changes independent of climate change. The shift to heavier isotopic values at Lubbock Lake occurred before the change in depositional environment, however, suggesting that the isotopic change reflects regionally driven environmental transitions in the biomass. The isotopic shift at the Lubbock Landfill also occurs within a uniform depositional environment as does that at Plainview, although much earlier.

The latest Pleistocene/early Holocene change toward isotopically heavier biomass appears to vary regionally. The change occurred earliest in the northern end of the study area (Plainview)

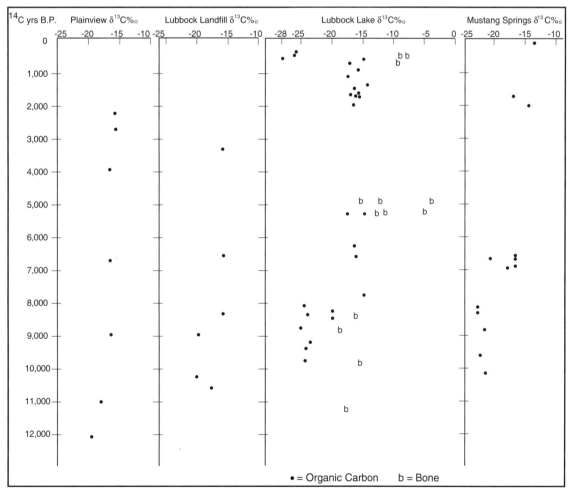

Figure 31. Plots of carbon isotope values from organic matter and bone versus age for the Plainview site, Lubbock Landfill, Lubbock Lake site, and Mustang Springs site (Fig. 4).

V. T. Holliday

at 12,000–11,000 yr B.P. and latest in the southern end of the area (Mustang Springs) at 8,000–6,500 yr B.P. Between these times (8,500–7,500 yr B.P.) the change occurred in about the middle of the area (Lubbock Lake). This slim data set hints at a time-transgressive environmental change: drying from north to south during the early Holocene.

The isotope data from the middle and late Holocene up to 500 yr B.P. are indicative of continued dryland vegetation, although there are only a few data points for the period 5,000–2,000 yr B.P. The change back to lighter $\delta^{13}C$ values in the past few hundred years at Lubbock Lake is indicative of an environmental change favoring more mesic flora. This change coincides with an increase in the distribution of stratum 5m, which may be linked to environmental changes that support increased spring discharge (Holliday, 1985c).

Ostracodes

by Vance T. Holliday

Ostracodes are microscopic crustaceans having a bivalved carapace composed of calcite. These organisms live in oxygenated surface and ground waters, from which hundreds of species are known for North America. Ostracode species assemblages may be variously used to describe paleohydrology and, in the case of wetland and lacustrine taxa, paleoclimate. Ostracode biogeographic ranges rapidly expand or contract (months to years), so as local or regional aquatic environments change that change is readily identified by changes in the ostracode fauna. These environmentally sensitive organisms have species that are often restricted to particular hydrological settings such as springs, streams, lakes, wetlands, or aquifers and then are further limited to particular ranges of physical and chemical properties. Thus, some or all of the ostracodes found in a cold, freshwater spring will differ from those in a cold freshwater lake, or those from a warm saline spring (Forester, 1991a). Moreover, taxa living in an aquifer, though poorly known ecologically, are very different in shell form from those living in any surface-water setting.

Lacustrine and wetland species' spatial distribution (biogeography) is limited by water temperature, major-dissolved ion (solute) composition, and concentration (total dissolved solids, TDS), all of which can be related to climate (Forester, 1987; DeDeckker and Forester, 1988). Seasonal stability of an environment may also be an important biogeographical factor (Forester 1991b). Because these species biogeographical limiting properties are the same ones that couple surface water to climate, an ostracode species' "distributional ecology" is also coupled to climate. Consequently, wetland and lacustrine ostracode species assemblages provide paleoclimate proxy records.

Quaternary ostracode studies were rare on the Southern High Plains prior to 1988. Given the abundance of lacustrine and palustrine sediments in the draws, however, the likelihood of ostracode preservation and recovery seemed high, and a pilot study of ostracodes in the valley fill was carried out by R. M.

Forester (personal communication, 1994). Ostracodes were extracted from only a few samples of stratum 3c (Table 23) in order to determine if they were present and if so to assess the environmental implications of these selected assemblages. The general paleoenvironmental implications of the ostracodes from stratum 3c follows.

In Running Water Draw, the Edmonson site was a freshwater (in composition and concentration) wetland to shallow lake. The TDS range likely was about 100 to 1,000 mg/l and the water was likely dominated by calcium and bicarbonate ions. The system may have been ephemeral on occasion and its physical and chemical properties varied somewhat seasonally. Of all of the sites examined, however, this was the most stable. The water

TABLE 23. OSTRACODES RECOVERED FROM STRATUM 3c AT VARIOUS SITES ON THE SOUTHERN HIGH PLAINS

Locality*	Sample Depth (cm)	Species
BLACKWATER DRAW		
Anderson Basin #1, Bw-71	0 - 18	No ostracodes recovered.
Gibson, Bw-31	75 - 100	*Cypridopsis vidua* *Potamocypris unicaudata* *Limnocythere staplin* also *Chara* spp.
	100 - 125	*Cypridopsis vidua* *Limnocythere staplin* also *Chara* spp.
MUSTANG DRAW		
Walker, Mu-17	180 - 230	*Cyprideis salebrosa*
EDMONSON, RW-19		
Edmonson, Rw-19	350 - 360	*Cypridopsis vidua* *Candona distincta* *C. inopinata* *C. compressa* *C. renoensis* *Paracandona euplectella* *Strandesia horridus* *Cyclocypris laevis*
Plains Paving, North Wall	123 - 134	No ostracodes recovered.
SULPHUR DRAW		
Brownfield, Su-16	430 - 455	*Cypridopsis vidua* *Candona compressa* *Lymnocythere* spp.
YELLOWHOUSE DRAW		
Enochs, Yh-33	274 - 300	*Candona rawsoni* southern version *Limnocythere ceriotuberosa* also *Chara* spp.
Lubbock Lake, Trench 73	200 - 210	*Candona rawsoni* southern version, mostly juveniles

*See Figure 4.

budget was supported by ground-water discharge and possibly some stream input. These wetlands were hydrologically open and may have had a topographic outlet and as a consequence occupied some intermediate position in the flow system. Biogeographic control is minimal, but limited data suggests that this assemblage is similar to those in the high valleys of Colorado, north-eastern Nevada, eastern Washington and Oregon, and parts of the Minnesota, Dakota, and Iowa prairies. This assemblage does not occur in modern lakes in Texas and Louisiana.

At the Gibson site, along Blackwater Draw, samples represent a slightly saline wetland that probably was near the terminus of the ground water flow path. *L. staplini* is a halobiont (a species that requires salt water to survive, grow, and reproduce) and can tolerate brine conditions, but the other taxa imply that the pond was at the low TDS end of the *L. staplini* range, somewhere within 1 to 5 g/l. This assemblage implies that the major ion composition evolved toward a calcium-enriched, bicarbonate-depleted water that likely was dominated by sodium and/or magnesium and sulfate and/or chloride. This water type is commonly the solute evolutionary product of water-sedimentary or granitic rock interactions. The environment likely varied seasonally and was ephemeral. Water dominated by calcium bicarbonates, typical of drinking water, is usually lethal to *L. staplini*, so it is unusual to find this taxon in a lake or wetland whose annual water budget was dominated by a stream. The *Chara* sp. implies ground-water discharge and/or spring discharge.

Ostracodes were recovered from two sites in Yellowhouse Draw, Lubbock Lake and Enochs, in addition to samples reported from Lubbock Lake by Pierce (1987). The ostracodes from stratum 3c at the Lubbock Lake site are likely reworked, but if not the water body was highly ephemeral and probably evaporated or lost water to seepage on a short-term basis (weeks). At Lubbock Lake, Pierce (1987, p. 41) reports "Ostracodes of the family Cyprididae . . . common to abundant in strata [1, 2, and 3], with *Candona* spp. observed most frequently." Possible identification of the species *Chlamydotheca* also was reported. Other species were observed in samples, but not reported. Identification of the genus *Candonna* provides little paleoenvironmental information because it is a ubiquitous taxa with a wide variety of species. Identification of *Chlamydotheca* seems unlikely because it is associated with tropical settings or warm spring-waters from deep aquifers (Forester, 1991a).

The Enochs site also may contain reworked ostracodes, but if not it was an ephemeral lake whose water budget was seasonally supported by stream discharge. Unlike the other samples this lake's major ion chemistry evolved towards calcium depletion and bicarbonate to carbonate enrichment or domination. High-carbonate, low-calcium waters are commonly the solute evolutionary by-product of water-volcanic rock interactions. The water in this system either interacted with volcanic ash along its flow path or removed calcium by some other mechanism such as the clay mineral exchange or precipitation as would occur with gypsum from a water with a large sulfate to calcium ration. The lake likely was seasonally saline.

Ostracodes are relatively rare at the Brownfield site in Sulphur Draw, so they may be reworked. Otherwise, this assemblage is common to ground-water-supported ephemeral wetlands. The ostracodes may have occupied a wetland located along a flow system rather than at the end of the flow system. TDS would typically be below about 2 g/l and could vary a great deal seasonally.'

Cuprideis salebrosa from the Walker site in Mustang Draw is an estuarine ostracode common to the Gulf of Mexico and Atlantic estuaries. It comes inland and is common in limnocrene (spring discharge into standing water pool) and rheocrene (spring discharge into a stream) settings having a water chemistry that is compatible, especially with regard to calcium versus bicarbonate, with the ocean. In most occurrence they are in waters that are dominated by sodium and chloride with high calcium and low bicarbonate, but they also occur in high calcium sulfate springs like Cuatro Ciénegas, Coahuila, Mexico, as well (Winsborough, 1990). The Walker site was a permanent spring having a TDS that could range from fresh water to values approaching normal marine salinity.

Although only a few samples were examined, the ostracodes recovered show promise for reconstructing the paleohydrology of sites such as this one. Paleoclimate reconstructions should also be possible, but would require that a large number of samples be examined in order to distinguish general climatically sensitive trends from local hydrological variability. The ostracodes show this region contained a wide range of paleohydrological settings that appeared, evolved, and disappeared from the landscape in response to climate change.

Molluscan remains

by Raymond W. Neck

Molluscan shells have been reported from late Quaternary strata on the Southern High Plains for many years (Pilsbry, 1935; Patrick, 1938; Neck, 1987; Pierce, 1987). However, most of these reports included analysis of shells from relatively few strata and generally from a single site. Environmental reconstructions were limited, usually indicating conditions similar to the present or conditions with significantly more moisture than at present. Few of these analyses discussed the relative role of meteorological and geomorphological controls on surface moisture conditions. This report is the beginning of a more comprehensive analysis of the paleomolluscan assemblages and the indicated environments of this area in latest Quaternary time. Samples were recovered from seven sites in Running Water Draw, Blackwater Draw, Seminole Draw, and Mustang Draw (Tables 24–30). Three Holocene sections sampled in this study and four sections reported previously illustrate general decreases in the availability of moisture on the Southern High Plains during the latest Pleistocene and early Holocene (Table 29).

The Plainview site samples (Table 25) provide an excellent temporal record documenting dramatic loss of species diversity from stratum 1 to stratum 2m (ca. 8,900 yr B.P.). However, some

TABLE 24. MOLLUSCAN REMAINS RECOVERED FROM VARIOUS SITES ON THE SOUTHERN HIGH PLAINS

Species		Sites*			
Age†	Carley Archer Stratum B Late Pleistocene	Seminole-Rose Stratum B 16.0	Eiland Stratum 1s 11.0?	Flagg Stratum 2d 8.5	Owen Ben Pit Stratum 4? Mid-Holocene?
Sphaerium striatinum					X
Pisidium nitidum				X	
P. casertanum			X		X
P. compressum			X		
Gyraulus crista					X
G. parvus			X	X	X
Micromenetus dilatatus			X		
Planorbella trivolvis	X				
Promenetus umbilicatellus				X	X
Laevapex fuscus			X		
Fossaria modicella					X
F. obrussa					X
Stagnicola caperata				X	X
S. elodes			X		
S. reflexa				X	
Physella virgata			X		X
Vallonia gracilicosta					X
Columella columella		X			X
Pupoides albilabris		X			X
Gastrocopta armifera					X
G. cristata					X
G. tappaniana					X
G. procera				X	
Vertigo ovata					X
Catinella avara					X
Succinea ovalis		X			
Helicodiscus singleyanus					X
Hawaiia minuscula					X
Euconulus trochulus					X
Number of species (29)	1	3	7	6	20

*See Figure 4.
†Ages in K yr B.P. based on radiocarbon chronology.

of this apparent loss of habitat could be the result of depositional environments providing different subsamples of the true fauna. A rich terrestrial environment is again indicated in stratum 3c. The apparent loss of terrestrial species diversity through time in stratum 3c probably is real as the depositional environments did not vary significantly.

Upper stratum 4 (the A horizon of the Lubbock Lake soil, postdating deposition of the stratum 4 sediments) appears to record an increase in available moisture relative to conditions during deposition of lower stratum 4 (eolian sediment was falling onto the marshy soil that capped stratum 3). The occurrence of *Discus cronkhitei* as late as upper stratum 4 seems to be rather late in the Holocene for a species that is presently unknown in Texas except for the relatively high elevations of the Guadalupe Mountains in far western Texas. This pattern probably reflects the ameliorated environments following early-middle Holocene drought in the region (see following section, "Summary, Discussion, and Conclusions").

Although the aquatic environment at Plainview also experienced a loss of diversity, this deterioration postdated the demise of most of the terrestrial fauna. The delayed loss of aquatic species diversity is probably the result of the overriding influence of ground water in an environment that was rapidly becoming drier. The freshwater gastropod record is interesting because of the rare occurrence of *Valvata tricarinata*, of which only a single shell was recovered from stratum 1gc. This operculate gastropod is an indicator of cold spring pools with water temperatures about 15° C. The stream may have had water velocities that were too rapid for this species or possibly the aquifer outflow was either dispersed along a series of seeps or was sufficiently upstream that only a few *V. tricarinata* shells were carried downstream by stream flow.

Two species previously unknown from the Southern High Plains were recovered from lower strata of the Plainview site: *Strophitus undulatus* (stratum 1s) and *Gastrocopta contracta* (stratum 2m-2). Other interesting species with a northern modern

TABLE 25. DISTRIBUTION OF MOLLUSCAN REMAINS RECOVERED FROM STRATA AT THE PLAINVIEW SITE*

Species Age**	Stratum†								
	1s	1gc	2m-1	2m-2	3c1	3cu	4s	4A	5A§
	9.0-10.0?		8.9		8.5"	7.0?	6.5	2.0	<1.0?
Strophitus undulatus	X		f						
Sphaerium striatinum	X	X	X				X	X	
S. simile	X							X	
Musculium transversum			X		X	X			
Pisidium nitidum	X								
P. casertanum								X	
P. compressum	X		X			X		X	
Valvata tricarinata		X							
Gyraulus parvus	X		X		X	X	X	X	
Promenetus umbilicatellus	X				X	X			
Fossaria cockerelli	X							X	
F. Dalli	X								
F. obrussa	X	X			X	X		X	
F. parva			X		X				
Stagnicola caperata	X		X						
S. elodes					X	X			
S. reflexa	X							X	
Physella virgata	X		X					X	
P. gyrina	X								
P. skinneri	X								
Aplexa hypnorum			X						
Carychium exiguum	X				X				
Vallonia parvula	X				X			X	
V. gracilicosta					X				
Columella columella					X				
Pupoides albilabris	X				X		X	X	
Gastrocopta armifera	X			X	X			X	
G. contracta				X					
G. cristata	X		X		X	X		X	
G. tappaniana	X				X	X		X	
G. pentodon	X		X						
G. pellucida					X				
G. procera	X			X	X	X	X	X	X
Vertigo milium			X		X	X	X	X	
V. ovata	X		X	X	X	X	X	X	
V. gouldi	X								
Strobilops texasianum					X				
Catinella avara		X	X		X	X	X	X	
Succinea ovalis	X								
Discus cronkhitei	X				X			X	
Helicodiscus singleyanus	X	X	X		X	X	X		
H. eigenmanni					X	X			
Deroceras laeve	X				X				
Hawaiia minuscula	X				X	X			
Zonitoides arboreus	X								
Number of species (45)	31	5	15	4	25	15	8	19	4

*Figures 4, 26. f = fragment of specimen.
†1s = sand and gravel; lgc = green clay within sand; 2m-1 = organic lamellae; 2m-2 = clay lamellae; 3cl = lower, 3cu = upper; 3s = A horizon at top of sand; 4s = middle; 4A = marshy, cumulic A horizon at top; 5 = A horizon (modern ground surface).
§Other unidentifiable shell remains of a *Sphaerium* sp. and two lymnaeid species were recovered from 5A.
**Ages in K yr B.P. based on radiocarbon chronology.

range present at Plainview but previously known from the late Quaternary of the Southern High Plains include the freshwater gastropods, *Physa skinneri* (seen only in stratum 1s) and *Aplexa hypnorum* (seen only in stratum 2m-1), both of which are typical of temporary habitats that may lose all surface water during the warm season. However, these species cannot survive in pond bottoms that become extremely dry and warm as do such sediments today in the modern Southern High Plains.

The paleoassemblages from the Wroe site (Table 26) indicate an asynchrony between the terrestrial and aquatic environmental changes. The overall ambient conditions at this site from about 10,500–8,300 yr B.P. vary as follows: a moist period, an arid period, a dry period, a wet period, and a moist period. The fluctuation of species composition also indicates changes in the water levels of ponds present during that time. Initially, there was a high water level, followed by a significant drop, then a return to a level near but not equal to the highest level. At the same time, terrestrial moisture initially was low, increased rapidly, fell once again, and then rose toward a level approximately equal to the previous peak. Most interestingly, the pond water level appears to vary inversely with terrestrial moisture. A decrease in moisture supply but retention of low evaporation levels appears to characterize the early Holocene period. Following is a period of increase in precipitation but also a significant increase in evaporation.

The most likely explanation of this pattern is the presence of a pond with variable water levels in a relatively constant volume depression. At times of low water level, more terrestrial habitat is available. The amount of moisture available for this terrestrial margin also varies through time. Overall, if the pond's water level is reduced, the terrestrial gastropods have a larger area to inhabit. In this example of asynchrony, the causal mechanism is believed to be intrinsic to geomorphologic constraints of the site, rather than only the gross climatic variations. Changes in water levels could be the result of variations in local precipitation runoff and concentration in a draw or changes in regional water tables. The early Holocene asynchrony at the Wroe site is much earlier than the asynchrony observed in samples from the Plainview site.

Samples from stratum 1 at Anderson Basin (Table 27) indicate the presence of a permanent pond with a well-drained upland and marsh habitat. Water level decreased and the upland marsh habitat was lost by stratum 2m time, as early as 10,700 yr B.P. Drake (1975) recovered freshwater gastropod species from strata B, 2d, and 2s. All these species are associated with temporary ponds and arid to semiarid woodlands, but with an increase of terrestrial habitat by ca. 10,000 yr B.P. and the beginning of the loss of both surface water and soil moisture by ca. 8,500 yr B.P.

Samples from the Gibson (Table 28) and Eiland sites also provide additional, although limited, paleoenvironmental insights. At Gibson, the 3s-1 and 3s-2 paleoassemblage is analogous, taxonomically, to shore dune field faunas along the Texas coast. These samples provide clues to the fauna of upland dune fields of the region. Very little is known of living faunas or the late Quaternary paleoassemblages in this setting.

One of the species present in the single Eiland site paleoassemblage, *Micromenetus dilatatus* (Table 24), is rather austral in distribution, ranging throughout the eastern United States as far west as Iowa, Missouri, and Texas. Most other aquatic species present in these paleoassemblages are more typical of the Great Plains today and more northern modern ranges. This species record may reflect a differential in lower-upper river basin zoogeographic connections between the Brazos and Colorado River systems during the late Quaternary.

Data from other molluscan surveys on the Southern High Plains (beyond those noted in Table 29) further document environmental changes through the Holocene. Pierce (1987) reported on molluscan paleoassemblages from Lubbock Lake. Molluscan remains from stratum 1 indicate the occurrence of a flowing stream with areas of quiet water and aquatic vegetation, and a substrate of sticks or stones for freshwater limpets. Subaerial habitats included hygric, marsh, and riparian woodlands with relatively little input from well-drained uplands. During the time of deposition of stratum 2 the aquatic resources were reduced to a pond and associated marsh with intermittent stream flows. The number of springs and amount of spring flow were greatly reduced as was the occurrence of riparian woodlands (both may have been absent). However, molluscan specimens from well-drained upland habitats were substantial. Lubbock Lake supported only temporary surface waters and all molluscan species with northern modern ranges were extirpated during the depositional time of stratum 3. Some marsh habitat was available, but extensive slope and well-drained upland habitats dominated the landscape. Conditions continued to desiccate during deposition

TABLE 26. MOLLUSCAN REMAINS RECOVERED FROM THE WROE SITE*

Species	Stratum†				
	1	2	3cl	3cm	3cu
Age§			10.1		8.3
Pisidium casertanum	X				
P. compressum	X				
Gyraulus parvus	X	X		X	
Planorbella tenuis	X	X		X	
Promenetus umbilicatellus		X		X	
Fossaria cockerelli				X	
F. dalli	X	X		X	X
Stagnicola caperata	X	X	X		
S. elodes		X			X
Physella virgata	X	X	X	X	X
P. gyrina	X	X	X	X	
Vertigo ovata			X		X
Oxyloma retusa			X		
Number of species (14)	8	8	5	7	4

*Figure 4.
†3cl = lower; 3cm = middle; 3cu = upper.
§Ages in K yr B.P. based on radiocarbon chronology.

**TABLE 27. MOLLUSCAN REMAINS RECOVERED
FROM ANDERSON BASIN 1***

Species			
	1	1/2	2m
Age§	11.0?		10.7
Pisidium casertanum	X		
Pisidium compressum	X		
Gyraulus crista	X	X	
G. parvus	X	X	X
Fossaria obrussa	X	X	X
Stagnicola elodes		X	X
Physella gyrina		X	X
Pupoides albilabris	X		X
Gastrocopta cristata	X		X
Vertigo gouldi	X		
Number of species (10)	8	5	6

*Figures 4, 10A, 23.
†Stratum 1/2 = mixed upper stratum 1 and lower stratum 2m.
§Ages in K yr B.P. based on radiocarbon chronology.

of stratum 4. Only small temporary ponds were present and there is no indication of marsh habitat. Isolated well-drained upland and protected slope habitats were occupied by gastropods. During the latest Holocene, as represented by stratum 5, probably only temporary ponding occurred, although surface water may have been more extensive than during the time of deposition of stratum 4. Terrestrial gastropods were limited to isolated protected areas with rock and vegetation present.

Paleoassemblages of fossil molluscs from stratum 2 at the Clovis site were reported by other investigators. All interpretations were essentially the same, including permanent, cool water with aquatic vegetation, wooded moist margins, and well-drained upland habitats (Pilsbry, 1935; Clarke, 1938; Wendorf, 1970; Drake, 1975). Clarke (1938) also reported species found in small streams and flood plains with extensive cover objects such as leaves or woody debris. Drake (1975, p. 208) proposed stratum 2 time as one of "general cool semiaridity in which considerable temporary ponding took place."

Gastropods from fill in lower Sulphur Springs Draw were recovered from strata 3s and 3c (strata 3 and 4, respectively, of Frederick, 1994; Neck, 1994). The shell from stratum 3s are representative of drought-tolerant taxa, but also include specimens of a freshwater species (*Fossaria cockerelli*). The setting was dry and well-drained, and surface water was available only in limited amounts. Gastropods from the marl are similar to those from 3s, but the assemblage lacks the freshwater taxa *Fossaria* sp. Species diversity of 3c is lower than 3s and may reflect an environment with greater soil alkalinity or salinity. The trend, therefore, is one of dry conditions getting drier.

At Mustang Springs, Fullington (in Meltzer and Collins, 1987) analyzed limited shell remains extracted from prehistoric water wells. Both terrestrial and freshwater gastropods were recovered from stratum 4, which filled the wells and buried stratum 3. These gastropods indicate the possible occurrence of still water and mesic marginal terrestrial habitats at this site at a time of the eolian deposition of stratum 4. An alternative hypothesis given by Meltzer and Collins (1987) is that these shells repre-

TABLE 28. MOLLUSCAN REMAINS RECOVERED FROM THE GIBSON RANCH SITE*

Species	Stratum†						
	2dl	2dm	2du	2m	3s-1	3s-2	3s-3
Age§		9.7				Early Holocene	
Pisidium casertanum	X	X		X			
Gyraulus crista	X	X	X				
G. parvus	X	X	X	X			
Promenetus umbilicatellus				X			
Fossaria cockerelli						X	
F. humilis					X		
F. obrussa	X	X	X				
Stagnicola elodes		X	X				
Physella virgata					X	X	X
P. gyrina		X	X				
Vertigo gouldi		X					
Catinella avara	X				X	X	
Oxyloma retusa	X						
Deroceras laeve						X	
Number of species (14)	6	7	5	3	3	4	1

*Figures 4, 29.
†2dl = lower, 307 cm; 2dm = middle, 300 cm; 2du = upper, 285-288 cm; 3s-1, 3s-2, 3s-3 = undifferentiated subsamples from organic-rich zones exposed in blowout.
§Ages in K yr B.P. based on radiocarbon chronology.

TABLE 29. PALEOECOLOGICAL SUMMARIES FOR MOLLUSC SAMPLES*

BLACKWATER DRAW

Anderson Basin 1, Bw-71 (Table 27)
Stratum 2m (ca. 10,700 yr B.P.):
 Isolated pond w/little or no marshy habitat; well-drained upland.
Stratum 1/2 (mixed upper 1 and lower 2):
 Pool or stream w/submerged vegetation.
Stratum 1 (ca. 11,000 yr B.P.?):
 Pool or stream w/submerged vegetation; bordered by marshy lowland and well-drained upland.

Clovis Site, Bw-59 and Bw-51 (Older Valley Fill) (Table 30)
Stratum B (<20,000 yr B.P.):
 Bcs — Temporary stream or variable-level lake w/extensive marshy habitat along margins; very mesic to hygric terrestrial habitats along margins of wetlands; no woody vegetation.
 Bss — Marsh w/seasonal ponds that regularly dessicate; well-drained, calcareous upland; limited indication of wetland-margin habitats; several species typical of protected (not necessarily hydric) and well-drained microhabitats.
Stratum A (ca. 20,000 yr B.P.):
 Shallow, low-volume, spring-fed stream w/some submerged vegetation; portions of stream possibly subterranean; terrestrial margin at least seasonally moist and supported very limited terrestrial fauna active under a low "canopy" of sedges and grasses.

Gibson Site, Bw-49, Bw-64, and blowout (Table 28)
Stratum 3s (early Holocene) (valley-margin blowout):
 3s-1 and 3s-2 — Temporary pond w/small, nonwoody-vegetation margin.
 3s-3 — Temporary pond w/no suggestion of terrestrial habitats suitable for terrestrial gastropod fauna.
Stratum 2m (<9,700 yr B.P.) Bw-49 (valley axis):
 Small stream or pool limited in area but probably permanent.
Stratum 2d (ca. 9,700 yr B.P. for upper sample) Bw-64 (valley axis):
 Upper 2d — Shallow pond w/permanent, cool water.
 Middle 2d — Shallow pond w/permanent, cool water, surrounded by marginal terrestrial marsh.
 Lower 2d — Shallow water pool w/some marsh margin present.

MUSTANG DRAW

Wroe Site, Mk-15 (Table 26)
Stratum 3c:
 Upper 3c (>5,700 yr B.P.) — Pond w/permanent core but variable in extent with limited mesic terrestrial margin.
 Middle 3c (ca. 8,300 yr B.P.) — Pond of variable size w/dependable, low-volume water supply, no indication of terrestrial margin suitable for gastropods.
 Lower 3c1 — Wetland margin microhabitats; pond subjected to extreme variations in water volume w/periodic desiccation; limited mesic to hydric terrestrial margin.
Stratum 2m (stratum 2/3 contact ca. 10,100 yr B.P.):
 Pond, probably permanent, but experienced more variation in site and depth than stratum 1 pond.
Stratum 1:
 Pond w/dependable water but highly variable in extent seasonally.

Eiland Site, Mk-12 (Table 24)
Stratum 1:
 Pond w/permanent water; submerged aquatic vegetation or dependable supply of wetland margin stems for substrate.

RUNNING WATER DRAW

Flagg Site, Rw-11 (Table 24)
Stratum 2d (<8,600 yr B.P.): Pond w/submerged vegetation; varied seasonally in extent; surrounded by either limited amount of terrestrial vegetation cover or, more likely, very limited drainage basin.

Plainview Site, Section 12, Pit 3 (Table 25)
Stratum 5s2 A horizon (1,000-0 yr B.P.):
 Substantial upland habitat w/xeric microhabitats able to support populations of only a few very hardy species; possible short-grass prairie; fragmented remains of aquatic species indicate scattered, possible intermittent ponds.
Stratum 4s (ca. 6,800-2,000 yr B.P.):
 Upper 4 (A horizon) — More diverse paleoassemblage; more mesic w/flowing stream; substantial marsh areas; upland habitat capable of supporting terrestrial gastropod populations.
 Lower 4 — Small stream with a narrow mesic margin and well-drained upland.

TABLE 29. PALEOECOLOGICAL SUMMARIES FOR MOLLUSC SAMPLES* (continued)

RUNNING WATER DRAW (continued)

Plainview Site, Section 12, Pit 3 (Table 25) (continued)

Stratum 3c (ca. 8,800-6,800 yr B.P.):

Upper 3 — Species diversity declines; stream or pool contained less water; terrestrial habitat probably had no significant amounts of woody vegetation.

Lower 3 — Small perennial stream w/associated pools of quiet water; lateral marsh habitat; mid-elevation riparian vegetation band w/woody plants including shrubs and some small trees; well-drained habitat upslope.

Stratum 2m (ca. 8,800 yr B.P.):

Small, slow-moving stream w/dependable water supply and submerged vegetation; very limited marsh area lateral to stream; clay lens within 2m associated with mesic to hydric terrestrial conditions.

Stratum 1 (>8,800 yr B.P.):

Permanent stream w/resident fish population and quiet water margins; lateral zone occupied by hydric marsh; abundant cover objects. Clay lens within stratum 1 associated w/cold water spring and pool w/variable water levels; terrestrial margin dry and harsh.

SEMINOLE DRAW

Seminole-Rose Site, Se-13 (Table 24)

Upper stratum B, A horizon (ca. 16,300 yr B.P.):

Pond margin habitat w/hygric soil conditions and well-drained uplands immediately upslope; protected microhabitats w/conservation of available soil moisture and reduced ambient evaporation.

*See Figure 4 for site locations.

---

sent secondary deposition of shells originally deposited in pond sediments of stratum 3. An interesting pattern in the stratigraphic distribution of these shells within stratum 4 is the occurrence of both aquatic species in all four subdivisions of stratum 4. In contrast, shells of terrestrial gastropods were rare in the lower part of stratum 4, most abundant in the middle portion (which constituted the upper portion of the fill from the well), but less abundant in the upper portion of stratum 4 that overlies the general surface of the location of the wells. Aquatic habitat was apparently present through this period, but terrestrial conditions were optimal for terrestrial gastropods only during the middle portion of this period.

Taken together, the fossil mollusc assemblages from the older valley and younger valley fill (Table 29) fill provide indications of local late Quaternary environments that may reflect regional conditions. The assemblages from strata A and B (20,000–15,000 yr B.P.) at the Clovis site (Table 30) and Seminole-Rose document the presence of ponds, marshes, and shallow spring-fed springs in the drainages late in the Pleistocene. The Clovis site fauna indicate active streams, marginal marshes, seasonal ponds, and mesic bank conditions. Despite the relative lushness of the environment indicated by the molluscs, no indication of woody vegetation is detected. The terrestrial species recovered from Seminole-Rose indicate the presence of a pond-margin habitat with mesic soil conditions; an area of well-drained uplands existed upslope from this pond margin. No aquatic species were recovered. This absence could result from taphonomic factors, sampling bias, or environmental conditions (e.g., intermittent surface water or high alkalinity) unsuitable for freshwater gastropods. The presence of *Succinea ovalis* and *Columella columella* indicate protected microhabitats with conservation of available soil moisture and reduced ambient evaporative factors, although the ambient

environment may have been very harsh for terrestrial gastropods (indicated by the lack of species diversity and the presence of a xeric-adapted species).

The fossil mollusc assemblages from stratum 1 (ca. 11,000–9,000 yr B.P.; Anderson Basin, Wroe, Eiland, Plainview) indicate a change in habitats available for molluscs (Table 29).

TABLE 30. MOLLUSCAN REMAINS RECOVERED FROM THE CLOVIS SITE*

Species	Stratum†		
	A	Bcs	Bss
Age§	>21.0	<21.0	
Pisidium nitidum		X	
P. casertanum	X		
P. compressum	X		
Gyraulus circumstriatus		X	
G. parvus	X		
Fossaria cockerelli			X
F. obrussa		X	
Carychium exiguum		X	
Vallonia parvula		X	
V. perspectiva		X	X
Pupilla muscorum		X	X
Pupoides albilabris			X
Gastrocopta cristata			X
G. tappaniana		X	X
Vertigo ovata		X	X
Catinella avara	X	X	X
Deroceras laeve		X	
Hawaiia minuscula		X	X
Number of species (18)	4	12	9

*Bcs and Bss from core Bw-58; A from core Bw-51; Figures 4, 12.

†Stratum Bcs = coarse sand in marl; Bss = sandy silt in marl.

§Ages in K yr B.P. based on radiocarbon chronology.

Anderson Basin at >10,700 yr B.P. contained a pond or stream with aquatic vegetation, a marshy margin, and a well-drained upland. The assemblage from mixed stratum 1/2 indicates a similar aquatic environment, but with significant loss of terrestrial habitat suitable for gastropods. The Wroe assemblage (>10,100 yr B.P.) indicates the presence of a pond with seasonally variable water levels, but no terrestrial species were recovered. The Plainview area during stratum 1 deposition (ca. 10,000–9,000 yr B.P.) contained a relatively moderate sized permanent stream and quiet backwaters, a hygric marsh with abundant cover objects, and a cold-water spring (near or upstream from the site). The single Eiland assemblage indicates the presence of a pond with emergent vegetation along the margins.

Fossil mollusc assemblages available from stratum 2 indicate further change in the habitats occupied by molluscs 11,000–8,000 yr B.P. (Flagg, Plainview, Anderson Basin, Gibson, Wroe; Table 29). Anderson Basin contained only an isolated pond with a limited marsh habitat by 10,700 (Table 27). The pond at the Wroe site at ca. 10,100 yr B.P. was still seasonably variable, but had a less dependable water supply (Table 26). The Gibson site supported a shallow pond with some marsh habitat; the three assemblages available from this site indicate a significant loss of marsh habitat by the time of deposition of the upper sample (probably between 9,500 and 9,000 yr B.P.; Table 28). The stream at Plainview (ca. 8,900 yr B.P.) was reduced and was flanked by a much more limited area of marsh habitat than existed previously (Table 25). Flagg at ca. 8,500 yr B.P. was a pond with submerged vegetation and arid terrestrial microhabitats (Table 24).

Fossil mollusc assemblages from stratum 3 (9,500–6,300 yr B.P.; Plainview, Wroe, Gibson) again demonstrate changes in the habitats available to molluscs, but the changes are not as unilateral as indicated in previous strata (Table 29). The lower assemblage (stratum 3cl) from Plainview (<8,900 yr B.P.) indicates the occurrence of a stream, marsh, riparian vegetation (including woody vegetation), and a well-drained upland. The upper assemblage from stratum 3 at Plainview (stratum 3cu; >6,800 yr B.P.) indicates significant aridification including the apparent loss of woody vegetation. Three assemblages from the Wroe site (8,300->5,700 yr B.P.) indicate the presence of a pond with variable water level and the terrestrial habitat suitable for gastropods varying from marsh to absent to semiarid with protected microenvironments. The Gibson site possessed a temporary pond with a sandy substrate and a semiwetland margin.

The Plainview site provides the only mollusc assemblages for stratum 4, but they provide significant information concerning the environments available at ca. 6,500 and 2,000 yr B.P. (Tables 25, 29). The older sample indicates the presence of a sluggish stream with a narrow margin of mesic habitat and a well-drained upland. The younger sample indicates a flowing stream with substantial marsh areas and riparian habitat. Most significant is the return of woody vegetation to this site during formation of the Lubbock Lake soil.

Only one assemblage from stratum 5 (Plainview 5A) of crushed freshwater gastropod shells and a single species of terrestrial gastropod is available (Tables 25, 29). Paleoenvironment conditions are difficult to interpret with this limited assemblage, but conditions for molluscs were probably quite harsh as indicated by the lack of shells rather than the identity of the single terrestrial species.

A conservative interpretation of the fossil mollusc assemblages through time reveals many variations in the types, dependability, and amount of habitat suitable for molluscs in the Southern High Plains in the latest Pleistocene and the Holocene (Table 29). At the close of the Pleistocene large ponds or lakes existed, but little information is available on the terrestrial environment (which may have been unfavorable for gastropods). Ambient climatic conditions were likely cool and dry to semiarid with low evaporation pressure. A period of landscape stability allowed some soil development and the development of mesic microhabitats suitable for terrestrial gastropods. Assemblages from stratum 1 indicate that the close of the Pleistocene was characterized by streams with marginal marshes and substantial suitable habitat for terrestrial gastropods. By the early Holocene, as represented by stratum 2, stream flow rates, pond water levels, and areas of marsh habitat were all significantly reduced. The molluscan fossils from the lower portion of stratum 3 indicate an increase in available moisture as streams and ponds increased in flow rates or water levels, but the assemblages from the upper portion of stratum 3 indicate a significant reduction in available water. Although the middle Holocene was dominated by eolian processes, stratum 4 assemblages indicate that a significant increase in available moisture occurred during the latter portion of this deposition, or at the end of significant eolian processes. In the early late Holocene woody vegetation returned to areas where these species were locally extirpated in the middle Holocene (upper stratum 3). This favorable time for molluscs was short lived as molluscan habitat rapidly deteriorated during the later late Holocene.

Several overall environmental inferences can be drawn from a general synthesis of the molluscan paleoassemblages recovered from sites on the Southern High Plains. Large perennial ponds or flowing streams existed in the latest Pleistocene and early Holocene at many locations along the draws; sedimentologic and stratigraphic data confirm the existence of flowing streams and ponds during this time period. In some localities, notably Plainview, this water supply was present throughout the Holocene, although generally in reduced amounts as time progressed. Formerly permanent ponds and streams became intermittent as the seasonal variation of the regional climate increased.

The only significant evidence of woody vegetation based on the mollusc species found in the assemblages was from the Ben site and in the middle and late Holocene assemblages at the Plainview site. However, this woody vegetation was most likely communities of shrubs and small trees similar in species composition to those present in very protected canyons along the Eastern Caprock Escarpment today. Timing of the environmental changes varies with site. Those sites with only very localized water sources were impacted in the early Holocene, whereas those with a regional stream that concentrated runoff from a large area or were near or downstream from a spring outlet (e.g., Plainview) retained a moderate water supply until the early middle Holocene and it returned in the late Holocene.

A lack of synchrony of deterioration of the aquatic and terrestrial environments was observed at several sites, but the timing and causal mechanisms of this asynchrony vary with site. Zoogeographically, the overall occurrence of species with northern modern ranges during the late Quaternary is accompanied by a lesser, but still significant, group of species from central and eastern Texas that indicate the existence of substantial supplies of surface and soil water at a time when temperatures were not critically lower than at present.

Several of these sites have revealed a promising paleoenvironmental record that was not fully explored during this study. The famous Clovis site has been studied for many decades, but a comprehensive paleoenvironmental record has yet to be presented for this site. Further analysis of a more complete set of samples from this site would begin to reveal the structural and temporal complexity and diversity of this locality. The Gibson site (including Marks Beach) revealed an apparent loss of habitat due to major environmental changes.

Diatoms

by Barbara M. Winsborough

Diatoms are useful in paleoecological analysis of lacustrine sediments because (1) in freshwater sediments diatoms tend to be the best preserved algal fossils, and (2) the most complete preservation of whole diatom communities is that of diatoms in Quaternary lake sediments, where probably 90% or more of the species that grew in the lake are preserved, along with a few species washed in from the catchment area (Round, 1981). These two characteristics of diatom assemblages are particularly well expressed on the Southern High Plains and diatomaceous sediments have been reported from the draws for almost a century (Tempere and Peragallo, 1907–1915; Lohman, 1936; Patrick, 1938; Hohn and Hellerman, 1961; Hohn, 1975; Winsborough, 1988; Meltzer, 1991).

During the 1988–1992 draw study, samples were taken from selected late Pleistocene and early Holocene strata at sites of marsh and pond deposition in Running Water Draw, Blackwater Draw, Yellowhouse Draw, and Mustang Draw (Tables 31, 32, 33). The most common species of diatoms found in the draws sediments are illustrated (Fig. 32). Species composition and paleoecological summaries of most sections are presented graphically (Figs. 33–36).

A total of 36 diatom species are recorded from the 6 samples at the Edmonson site, with 8 species reaching abundances of at least 10% of a population (Fig. 33). *Epithemia argus* is overwhelmingly the most common diatom in the section (Table 33). The diatoms suggest that the oldest sample represents a shallow,

muddy, saline interval during initial filling of a lake, followed by relatively deeper (1–2 m), less saline, more vegetated water conditions. This shift is reflected in a decrease in *Rhopalodia gibberula* and *Anomoeoneis* spp. taxa generally indicative of elevated salt levels when growing in abundance. Soil and mud dwellers decrease upsection, giving way to eophytes (Fig. 33). The upper half of the section may represent a gradual decline in water level, but no substantial increase in salinity.

At the Flagg site (Table 33; Fig. 33) the dominant diatom species are characteristic of a shallow, marshy, vegetated freshwater pond or lake with a slightly alkaline pH of about 7.5–8.5. Nutrient concentrations were probably low to moderate, and total dissolved solids and conductivity rather high. There are more soil and aerophilic (characteristic of nonsubmerged habitats such as moss, damp sand, rock, or mud) species in the lower part of the section and more epiphytic and epipelic (mud) species in the middle and upper part of the section. There is no evidence of a substantial current other than what might be produced by a flowing spring. The autecological characteristics of the dominant diatoms indicate a gradual deepening, or increase in size of the lake over the period represented (Fig. 33). This long-term change is most likely related to increased volume of spring flow. Within this overall trend there are reversals in the general pattern representing the response of diatoms to short-term changes in their immediate habitat. These changes could be brought about by flood events that scoured away the sediment and aquatic vegetation, or by drought conditions slowing the flow of springs and seeps and allowing the development of emergent vegetation.

In the samples from the Clovis site (Table 33; Fig. 34) all of the diatoms are benthic (bottom dwelling) species found on plants, and in sand and mud. Many of the species are aerophilic or soil species. These characteristic species suggest a spring or seep with flowing, well-aerated, shallow water. All of the common and abundant species prefer extremely hard, high-conductivity water, and most are characteristically found in alkaline, slightly to definitely saline water high in carbonate or chloride salts. The water temperature was rather warm at least part of the year, as over half of the dominant species prefer warm water, as expected in a shallow lake, where the diatoms are living in the littoral zone, close to the water surface. The lower part of the section represents a shallow-water environment. Subsequently there is a gradual deepening trend up section (Fig. 34). Depth may never have been more than a meter where these samples were collected. The middle of the section may represent a particularly dry interval, but the original trend resumes in the upper section.

These conclusions agree with earlier investigations at Clovis by Lohman (1936) and Patrick (1938). There are more soil and moss species in the present study, suggesting that a damp mossy cover over mud or rock probably existed at least part of the time, where these samples were collected. The sampling at Clovis in the 1930s was in the deeper areas of the paleobasin.

At the Davis site (Table 33) there are few lacustrine species and their abundance is very low. The assemblage is characteristic of a very shallow spring-fed marsh with a partially vegetated,

**TABLE 31. LIST OF DIATOM SAMPLE LOCATIONS, STRATA SAMPLED,
CHRONOLOGICAL RELATIONSHIPS, AND NUMBER OF SAMPLES ANALYZED**

Site*	Draw	Core or Trench Number	Stratum (Interval, cm)	Age (^{14}C yrs B.P.)	Number of Samples
Clovis†	Bw	South Bank	2d (#90-95) 2s (#95a)	10,740-10,110	7
Davis	Bw	Bw-45	2m (300-330)	10,900	1
Edmonson	Rw	Rw-19	3c (360-374) 2m (374-400)	9,800	7
Flagg	Rw	Rw-11	2d (255-283) 2m (283-300)	9,400-8,600	8
Gibson	Bw	Bw-31	2d (331-370)	9,700	40
Glendenning	Mu	Mu-3	4s (155-170) 3c (170-175)	9,500	5
Lubbock Lake	Yh	Tr 73	2m (237-249) 2d (249-264)	10,000-6,500 11,000-10,000	13
Lubbock Lake	Yh	Tr E-1	1m, 2d, 2m, 3c, 5m (0-253)	11,000-400	44
Mustang Springs	Mu	Mu-1	1, 2d, 2m, 3c (212-360)	10,200-100	74
Tolk	Bw	Bw-36	2m (180-194)	10,100	8

*Figure 4.
†Sample intervals are sample numbers from Haynes, 1995, Table 2. Samples 90-95 are Unit D_1 of Haynes, 1995 (stratum 2d in this volume) and sample 95a is from Unit D_0 of Haynes, 1995 (stratum 2s in this volume).

muddy bottom, an alkaline pH, and high total dissolved solids, particularly carbonates and sulfates. The marsh probably was subjected to periodic or episodic desiccation.

There are two noteworthy characteristics of the diatom assemblage from the Gibson site (Table 33; Fig. 34). The first is that the species diversity declines gradually from bottom to top of the section, as shown by the number of species in each sample (Fig. 34). This change may indicate an environmental shift such as gradual loss or reduction of habitat or substrate associated with fluctuations in water depth or salinity. The second characteristic is that among the commonly occurring species, there are two sets of diatoms that alternate with each other in abundance, that is, when one group is abundant the other is scarce. Set #1

includes *Nitzschia tropica, Epithemia argus, E. turgida, Synedra capitata,* and *Melosira italica.* These species alternate with set #2, *Synedra ulna* and *Cocconeis placentula.* Set #1 suggests relatively stable, vegetated lacustrine conditions compared to the second set, which represents habitat in the early stages of colonization, with limited plant and mud substrate, favoring opportunistic species. Water levels fluctuated through time, from relatively deep water (no more than a few meters) to shallow water or no standing water. Gradually, seasonally fluctuating marshy conditions predominated (Fig. 34).

The Tolk site diatoms (Table 33) are all benthic forms living on vegetation and sediment. They indicate alkaline, high-conductivity fresh to slightly saline, shallow, probably

TABLE 32. SPECIES LIST AND ALTERNATE NAMES* FOR LATE QUATERNARY DIATOMS FROM THE SOUTHERN HIGH PLAINS

Achnanthes affinis Grunow
A. exigua Grunow
A. gibberula Grunow
A. lanceolata (brébisson) Grunow
A. lanceolata var. *dubia* Grunow
A. microcephala (Kützing) Grunow
Amphora coffeaeformis (Agardh) Kützing
A. holsatica Hustedt
A. ovalis (Kützing) Kützing
A. perpusilla Grunow
A. sabiniana Reimer
A. veneta Kützing
Anomoeoneis costata (Kützing) Hustedt
   A. sphaerophora f. *costata* (Kützing) Schmidt
A. sphaerophora (Ehrenberg) Pfitzer
A. sphaerophora var. *guentherii* O. Müller
A. vitrea (Grunow) Ross
Caloneis amphisbaena (Bory) Cleve
C. bacillum (Grunow) Cleve
C. limosa Patrick and Reimer
   Caloneis schumanniana (Grunow) Cleve
C. schumanniana (Grunow) Cleve
C. ventricosa (Ehrenberg) Meister
C. hibernicus Ehrenberg
Cocconeis pediculus Ehrenberg
C. placentula Ehrenberg
Cyclotella glomerata Bachmann
C. meneghiniana Kützing
Cymatopleura elliptica (Brébisson) W. Smith
C. solea (Brébisson) W. Smith
Cymbella aspera (Ehrenberg) Cleve
C. cistula (Ehrenberg) Kirchner
C. cymbiformis Agardh
C. delicatula Kützing
C. lata Grunow
C. hauckii Van Heurck
C. mexicana (Ehrenberg) Cleve
C. microcephala Grunow
C. minuta Hilse
C. minuta var. *pseudogracilis* (Cholnoky) Reimer
C. norvegica Grunow
C. pusilla Grunow
Denticula elegans Kützing
D. kuetzingii Grunow
D. subtilis Grunow
Diploneis elliptica (Kützing) Cleve
D. oblongella (Naegeli) Cleve-Euler
D. pseudovalis Hustedt
Epithemia adnata (Kützing) Brébisson
E. argus Grunow
E. turgida (Ehrenberg) Kützing
Eunotia curvata (Kützing) Lagerstedt
E. monodon Ehrenberg
E. pectinalis (O. F. Müller?) Rabenhorst
E. pectinalis var. *minor* (Kützing) Rabenhorst
Fragilaria brevistriata Grunow
   Pseudostaurosira brevistriata (Grunow) Williams and Round
F. capucina Desmazières
   Synedra rumpens Kützing
F. construens (Ehrenberg) Grunow
   Staurosira construens Ehrenberg
F. construens var. *binodus* (Ehrenberg) Grunow
F. elliptica Schumann
F. fasciculata (Agardh) Lange-Bertalot
F. leptostauron (Ehrenberg) Hustedt

Staurosirella leptostauron (Ehrenberg) Williams and Round
F. nitida Héribaud
F. nitzschioides Grunow
   Neofragilaria nitzschioides Grunow
F. pinnata Ehrenberg
   Staurosirella pinnata (Ehrenberg) Williams and Round
F. pseudoconstruens Marciniak
   Pseudostaurosira pseudo. (Marciniak) Williams and Round
F. robusta (Fusey) Manguin
F. vaucheriae (Kützing) Petersen
   F. capucina var. *vaucheriae* (Kützing) Lange-Bertalot
Gomphonema acuminatum Ehrenberg
G. affine Kützing
G. angutum (Kützing) Rabenhorst
   G. intricatum Kützing
   G. intricatum var. *vibrio* (Ehrenberg) Cleve
   G. dichotomum Kützing
G. brebissonii Kützing
G. clavatum Ehrenberg
   G. subclavatum (Grunow) Grunow
   G. montanum Schumann
G. gracile Ehrenberg emen. Van Heurck
G. insigne Gregory
   Gomphonema gracile Patrick and Reimer
G. parvulum (Kützing) Grunow
G. truncatum Ehrenberg
Hantzschia amphioxys (Ehrenberg) Grunow
H. vivax (W. Smith) Peragallo
Mastogloia elliptica var. *dansei* (Thwaites) Cleve
M. grevillei W. Smith
M. smithii var. *lacustris* Grunow
Melosira italica (Ehrenberg) Kützing
   Aulacoseira italica (Ehrenberg) Simonsen
Meridion circulare (Greville) C. A. Agardh
Navicula accomoda Hustedt
N. amphibola Cleve
N. angusta Grunow
N. auriculata Hustedt
N. capitata Ehrenberg
N. capitata var. *hungarica* (Grunow) Ross
N. cryptocephala Kützing
N. cuspidata var. *heribaudii* M. Peragallo
N. elginensis var. *lata* (M. Peragallo) Patrick
N. elginensis var. *rostrata* (A. Mayer) Patrick
N. erifuga Lange-Bertalot
   N. heufleri var. *leptocephala* (Brébisson) Peragallo
N. gallica (W. Smith) Lagerstedt
N. gastrum (Ehrenberg) Kützing
N. halophila (Grunow) Cleve
N. heufleri Grunow
   N. cincta (Ehrenberg) Ralfs
N. integra (W. Smith) Ralfs
N. laevissima Kützing
N. lubbockii Hohn and Hellerman
N. mutica Kützing
N. oblonga Kützing
N. peregrina (Ehrenberg) Kützing
N. platensis (Frenguli) Cholnoky
   N. braziliana Cleve
N. pupula Kützing
N. pygmaea Kützing
N. radiosa Kützing

N. recens (Lange-Bertalot) Lange-Bertalot
N. schroeterii Meister
N. texana Patrick
   N. kotschyi Grunow
N. viridula var. *rostellata* (Kützing) Cleve
Neidium affine (Ehrenberg) Pfitzer
Neidium ampliatum (Ehrenberg) Krammer
N. bisulcatum var. *baikalense* Skvortzow and Meyer
N. iridis (Ehrenberg) Cleve
Nitzschia amphibia Grunow
N. angustata Grunow
N. brevissima Grunow
N. compressa (Bailey) Boyer
   N. punctata (W. Smith) Grunow
N. constricta (Kützing) Ralfs
   N. apiculata (Gregory) Grunow
N. debilis Arnott
   N. tryblionella var. *debilis* (Arnott) Hustedt
N. dubia Wm. Smith
N. fonticola Grunow
N. frustulum (Kützing) Grunow
N. kittlii Grunow
N. microcephala Grunow
N. palea (Kützing) W. Smith
N. punctata (W. Smith) Grunow
N. tropica Hustedt
N. tryblionella Hantzsch
N. vitrea Norman
Opephora martyi Héribaud
Pinnularia abaujensis (Pantocsek) Ross
P. acrosphaeria Rabenhorst
P. appendiculata (Agardh) Cleve
P. borealis Ehrenberg
P. microstauron (Ehrenberg) Cleve
P. nodosa Ehrenberg
P. subcapitata Gregory
P. viridis (Nitzsch) Ehrenberg
Rhoicosphenia curvata (Kützing) Grunow
Rhopalodia gibba (Ehrenberg) O. Müller
R. gibberula (Ehrenberg) O. Müller
R. musculus (Kützing) O. Müller
R. parallela (Grunow) O. Müller
   R. gibba var. *parallela* (Grunow) H. and M. Peragallo
Stauroneis acuta var. *terryana* Tempere ex Cleve
S. anceps Ehrenberg
S. phoenicentron (Nitzsch) Ehrenberg
S. pseudosubobtusoides Germain
S. smithii Grunow
Surirella angusta Kützing
S. brebissonii Krammer and Lange-Bertalot
S. hoefleri Hustedt
S. linearis W. Smith
S. ovalis Brébisson
S. ovata Kützing
S. spiralis Kützing
Synedra capitata Ehrenberg
   Fragilaria dilatata (Bréisson) Lange-Bertalot
S. parasitica (W. Smith) Hustedt
   Fragilaria parasitica (W. Smith) Grunow
S. ulna (Nitzsch) Ehrenberg
   Fragilaria ulna (Nitzsch) Lange-Bretalot
S. ulna var. *acus* (Kützing) Lange-Bertalot
   Fragilaria acus (Kützing) Lange-Bertalot
   S. acus Kützing

*Alternate names are indented.

TABLE 33. DISTRIBUTION AND RELATIVE ABUNDANCE* OF LATE QUATERNARY DIATOMS ON THE SOUTHERN HIGH PLAINS

Species	Site†									
	1	2	3	4	5	6	7	8	9	10
Achnanthes affinis								R		
A. exigua	R		O					R		O
A. gibberula										O
A. lanceolata										C
A. lanceolata var. *dubia*	R	R	O		R					O
A. microcephala		R	R		O	R		R		A
Amphora coffeaeformis	O				R			R	O	R
A. holsatica										R
A. ovalis	C	O	O		C	R	O	C		O
A. perpusilla		O	R		O					A
A. sabiniana		R			R					O
A. veneta		O	O		R					O
Anomoeoneis costata		R					A			O
A. sphaeophora	R	R	O		O	O	A	O		R
A. sphaerophora var. *guntherii*	R	O		R		O	O	R		O
A. vitrea										R
Caloneis amphisbaena					R					
C. bacillum	O	O	O		O		R	O	R	R
C. limosa			C		O					O
C. schumanianna		R								
C. ventricosa	R				R		R	R		
C. hibernicus		R	R					O		
Cocconeis pediculus								R		
C. placentula	O	O			A		R	O	R	A
Cyclotella glomerata								R		
C. meneghiniana	R	O			O			R		O
Cymatopleura elliptica		R								R
C. solea					O					R
Cymbella aspera										O
C. cistula	R	O			O	O	R	O		A
C. cymbiformis					O	O		O		A
C. delicatula	R									C
C. lata								R		
C. hauckii										O
C. mexicana				R	R		O	A		O
C. microcephala	R				R					
C. minuta			O		O		R	R		R
C. minuta var. *pseudogracilis*		R								
C. norvegica			R		O					
C. pusilla		R			R					
Denticula elegans	A	A	A	A	C	O	A	C	A	A
D. kuetzingii		R			O					A
D. subtilis										O
Diploneis elliptica	O	O	O		O		O	A	R	O
D. oblongella			O		R					
D. pseudovalis	O	R	C		R			C	O	O
Epithemia adnata		O	A		A			R		R
E. argus	A	A	C	C	A	A	A	A	C	C
E. turgida	A	R	O	R	A	A	A	A	O	C
Eunotia curvata		R			O					R
E. maior	O				C					O
E. pectinalis					O			O		
E. pectinalis var. *minor*								R		
Fragilaria brevistriata	A	A			C			O		A
F. capucina					R			O		C
F. construens		R						C		C
F. construens var. *binodus*	R	O								O
F. elliptica										O

TABLE 33. DISTRIBUTION AND RELATIVE ABUNDANCE* OF LATE QUATERNARY DIATOMS ON THE SOUTHERN HIGH PLAINS (continued)

Species	Site†									
	1	2	3	4	5	6	7	8	9	10
F. faciculata										R
F. leptostauron	R									
F. nitida		O			O					A
F. nitzschioides	C	O	O			O	C	A	O	A
F. pinnata	O				O		A	C		O
F. pseudoconstruens										O
F. robusta										O
F. vaucheriae					O					R
F. virescens	C		O			O	C	A	O	A
Gomphonema acuminatum	R	R			O			O		O
G. affine	O	R	O	R	O	C	C	C	O	A
G. angustum	R	O			O					C
G. brebissonii					O					
G. clavatum					O					
G. gracile										R
G. insigne										R
G. parvulum		R			O			R		O
G. truncatum		R			O					
Hantzschia amphioxys		R	C	R	O			O	R	O
H. vivax		O		O	O	O	C	O	R	O
Mastogloia elliptica var. *dansei*	O	O			O					O
M. grevillei					R					
M. smithii var. *lacustris*		O			O			R	R	O
Melosira italica	C	A			A	A	C	C	O	O
Meridion circulare	R	R						O		O
Navicula accomoda	R	R								
N. amphibola	R				R		R			
N. angusta		O			O					
N. auriculata										R
N. capitata					O					R
N. capitata var. *hungarica*								R		R
N. cryptocephala		C			C					O
N. cuspidata var. *heribaudii*	O	O	O	R	O	O	A	O		R
N. elginensis var. *lata*								R		
N. elginensis var. *rostrata*	O	R	O		O			O		O
N. erifuga										O
N. gallica			A		O					A
N. gastrum								R		
N. halophila					O					O
N. heufleri	C	R	O		C		R	R		O
N. integra										O
N. laevissima					R					
N. lubbockii					O					
N. mutica		R	O		R					R
N. oblonga	O	C			C	O	C	O		C
N. peregrina					R					
N. platensis	R	R								R
N. pupula	R	R	O		O		R			O
N. pygmaea		R			O					O
N. radiosa					R					
N. recens		O			O					R
N. schroeterii										R
N. texana		R			R					R
N. viridula var. *rostellata*								R		O
Neidium affine					R					
Neidium ampliatum	R	O			R					O
N. bisulcatum var. *baikalense*	O			O	O	O	R			
N. iridis					R					

TABLE 33. DISTRIBUTION AND RELATIVE ABUNDANCE* OF LATE QUATERNARY DIATOMS ON THE SOUTHERN HIGH PLAINS (continued)

Species	1	2	3	4	5	6	7	8	9	10
Nitzschia amphibia	C	A	A	R	A		O	O	C	A
N. angustata					O					
N. brevissima	R									
N. compressa					O					
N. constricta		A								
N. debilis			R		O					O
N. dubia	O	R								
N. fonticola					R					
N. frustulum		A			O		O			O
N. kittlii			R		O					O
N. microcephala			O		O					O
N. palea										R
N. punctata								R		
N. tropica	O	O	O		A					O
N. tryblionella		C								R
N. vitrea		R			O					R
Opephora martyi		R			O			R		
Pinnularia abaujensis					O			O		
P. acrosphaeria		R			R		R			
P. appendiculata		C			R					O
P. borealis		R	R	O	O					
P. microstauron	O	O	C	O	O	O	C	O	O	R
P. nodosa			O		R					
P. subcapitata	O	R	R		O		C	R	R	O
P. viridis	O	O	O	O	O	C	C	A	O	O
Rhoicosphenia curvata										R
Rhopalodia gibba	A	A	A	O	A	O	A	A	C	A
R. gibberula	A	C	A	A	C	O	C	A	A	A
R. musculus										O
R. parallela										O
Stauroneis acuta var. *terryana*			R					O		O
S. anceps					R					R
S. phoenicentron					O	R		O		
S. pseudosubobtusoides		O	O		O					O
S. smithii	R		O							O
Surirella angusta					R					R
S. brebissonii					O					
S. hoefleri					O					
S. linearis			O							
S. ovalis					O		R			
S. ovata					R					
S. spiralis		R	R		O					R
Synedra capitata		O			A	R	O	O		O
S. parasitica		A	O							C
S. ulna	C	O		R	A	A	C	C	C	A
S. ulna var. *acus*										C

*Abundance (data based on a count of 500 cells in each sample); A = Abundant (at least 10 percent relative abundance in any sample; i.e., ≥50 individuals); C = Common (5-10 percent of any sample, i.e., 25-49 individuals); O = Occasional (1-5 percent of any sample, i.e., 5-24 individuals); R = Rare (less than 1 percent of any sample, i.e., 1-4 individuals).

†Site key (see Figure 4 for site locations, see Table 31 for sampled strata and age control); Site 1 = Lubbock Lake, Tr 73; Site 2 = Lubbock Lake, Tr E-1; Site 3 - Clovis, South Bank; Site 4 = Davis, Bw-45; Site 5 = Gibson, Bw-31; Site 6 = Tolk, Bw-36; Site 7 = Edmonson, Rw-19; Site 8 = Flagg, Rw-11; Site 9 = Glendenning, Mu-3; Site 10 = Mustang Springs, Mu-1.

nutrient-rich water. There was probably a combination of standing and flowing water.

Two sets of samples were analyzed from Lubbock Lake. Thirteen lithologically distinct horizons were sampled from Trench 73 (Figs. 28A, 35) and 44 samples are analyzed from Trench E-1 (Fig. 28A). Both sections were sampled at regular intervals (Table 31; Fig. 35). Trench 73 is in the valley axis near the central part of the paleolake. Core E-1 is closer to the valley margin, representing less stable habitat. Trench 73 diatoms show a sequence of alternating marsh and lake conditions during the earliest Holocene. Many of the samples in core E-1, although diatomaceous, were too heavily diluted in sediment and organic matter to obtain reliable quantitative data. Reworking of older sediments appears to have been common in this core. This interpretation is supported by the location of the core near the valley margin.

This study essentially confirms the interpretations of Hohn and Hellerman (1961), based on their work at Lubbock Lake. In the present study, several very short term shifts between shallow marsh conditions and lake conditions, ending with the same saline (brackish), drying conditions, were detected due to closer sampling intervals (Table 33; Fig. 35). The paleolake was circumneutral to at least slightly alkaline, with a pH of about 7 to 8.5. Under high water table lacustrine conditions, the total dissolved solids and conductivity were moderate. Under drier, low-flow water table conditions, the diatoms suggest a water chemistry characterized by high conductivity and high total dissolved solids, typical of the evaporative setting found in a marsh with limited or seasonal flow. There is no evidence of eutrophic conditions.

Hohn and Hellerman (1961) attributed the changing diatom assemblages at Lubbock Lake and Clovis to regional climatic changes. General trends in the assemblages may be due to regional factors, but individual fluctuations, between relatively deep and shallow water conditions, could result from local influences. Moreover, because of microtopographic variation along the floor of the draw (especially normal to the trend of the draw), there might be ponds in some places and, not far away, marshy ground with water at or below ground level. Therefore, samples from different places in the draw might show different types of assemblages of the same age. Core E-1 exemplifies these differences. Although the section is long, many of the samples were substantially diluted with sediment and contain a mixture of soil and aquatic species.

Only one sample from the Glendenning site was diatomaceous. The diatoms are soil or marsh varieties and indicate dry or fluctuating wet and dry conditions (Table 33). There also is some indication of saline conditions occurring seasonally.

The samples from Mustang Springs (Table 33; Fig. 36) compliment results of analyses reported elsewhere (Winsborough 1988; Meltzer, 1991). The site evolved from a flowing spring or stream with well-aerated, alkaline water to a shallow, marshy to marly evaporitic setting, with rather stagnant water, and usually with abundant aquatic vegetation and varying water depth and salinity. There were two deep-water intervals separated by shallow muddy or marshy conditions. Lacustrine conditions never were reestablished after 6,500 yr B.P., although diatoms, particularly soil species, are present though sparse throughout the upper part of the section.

The late Pleistocene and early Holocene diatoms found in the lakes and marshes of the draws are dominated by a single overall association. Of the total of 161 species of diatoms recorded from the sites (Table 32) 31 occur in abundant numbers in at least one sample (Table 33). These species are characteristic of a pH neutral to alkaline, high-conductivity fresh water to slightly saline, spring-fed marsh to marshy lake habitat, with a fluctuating water table. They are benthic or opportunistically planktonic species. A true plankton population never developed at these sites. The shallow, clear water favored benthic growth. The most abundant and widely distributed diatoms on the Southern High Plains during the late Quaternary are *Denticula elegans*, *Epithemia adnata*, *E. argus*, *E. turgida*, *Fragilaria virescens*, *F. brevistriata*, *Melosira italica*, *Nitzschia amphibia*, *Rhopalodia gibberula*, *R. gibba*, and *Synedra ulna*. *D. elegans* is the most widespread diatom, being abundant at seven of the ten sites.

One species, *R. gibberula*, is the most salt-tolerant species that flourished on the Southern High Plains. It grows well with salinities up to 70 ppt. and is considered to be indicative of saline (brackish) conditions. Its presence with less salt tolerant species indicates that the water was usually not more than slightly saline. The other diatoms are oligohalobous (freshwater forms occuring in salt concentrations of less than 500 mg/ml), or halophilous (growing best with small concentrations of carbonate or chloride salts). They are indifferent to current, and are for the most part eurythermal (occuring over a temperature range of 15° C or better), but a few species indicate cool conditions. Some of the abundant species live in damp soil and moss, but this is compatible with the autecological summary given above and illustrates the range of associated environments. With regard to pollution, the diatoms indicate low to moderate nutrient and pollutant levels. According to Krammer and Lange-Bertalot (1991), one species typically found in spring-fed ponds and ditches, *Fragilaria virescens*, tends to come and go due to increasing eutrophication found near civilization–"*zivilizationsnaher*" (human-impacted settings or sites). This species is abundant at Flagg and Mustang Springs, and occasional to common at five other sites.

A shallow water body is heated considerably in the summer, so species living in the littoral zone would have to be conditioned to warm water. The cool-water forms are found in abundance today in northern, subboreal lakes and marshes, and may indicate a late Pleistocene/early Holocene winter air temperature slightly lower than at present. These conditions may have produced slightly lower ground-water temperatures. Through time the diatom assemblages from the draws show decreasing diversity and a selection for robust forms such as *Epithemia*, *Deticula*, and *Rhopalodia*, with internal chambers to aid in adjustment to

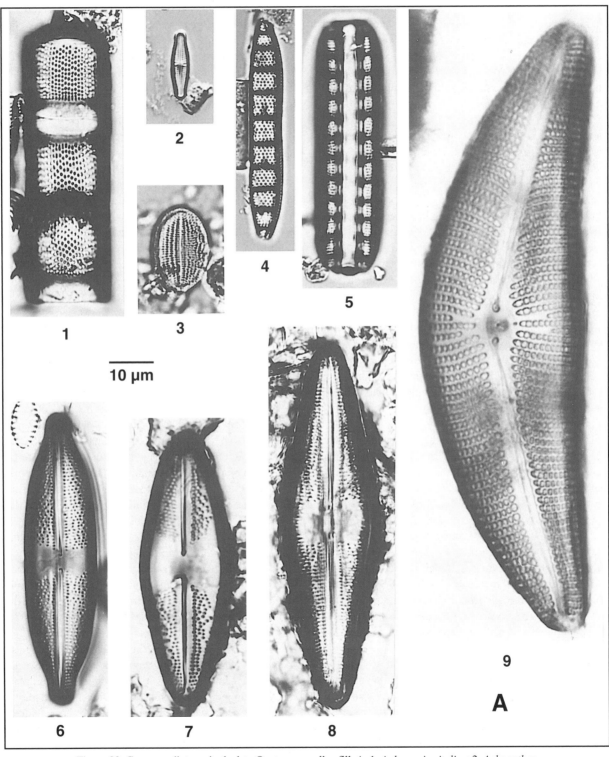

Figure 32. Common diatoms in the late Quaternary valley fill. A. 1, *Aulacoseira italica*; 2, *Achnanthes microcephala*; 3, *Cocconeis placentula*; 4, 5, *Denticula elegans*; 6, *Anomoeoneis sphaerophora*; 7, 8, *Anomoeoneis costata*; 9, *Cymbella mexicana*.

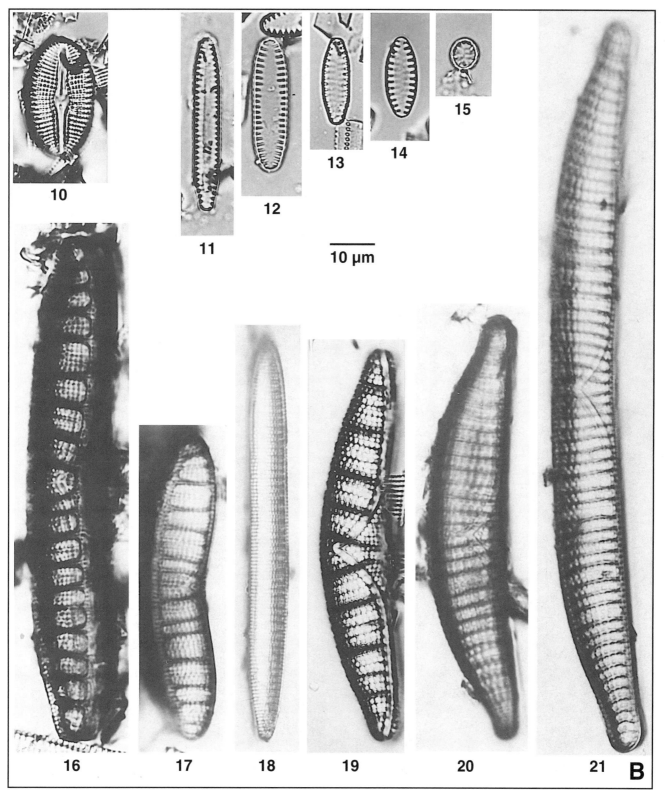

Figure 32B. 10, *Diploneis elliptica*; 11–15, *Fragilaria brevistrata*, illustrating size range; 16, 17, *Epithemia adnata*; 18, *Denticula kuetzingii*; 19, *Epithemia argus*; 20, 21, *Epithemia turgidda*.

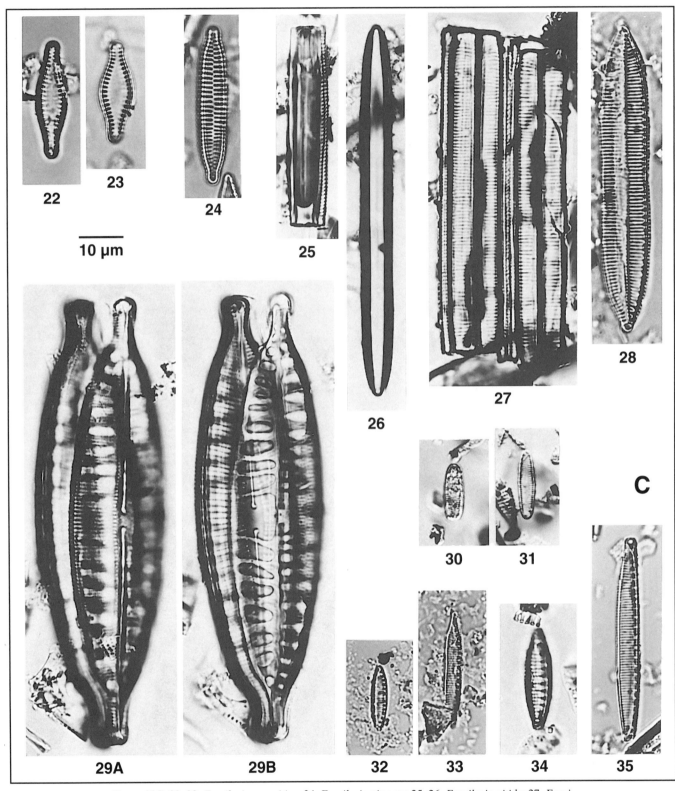

Figure 32C. 22, 23, *Fragilaria parasitica*; 24, *Fragilaria pinnata*; 25, 26, *Fragilaria nitida*; 27, *Fragilaria nitzschioides*; 28, *Nitzschia constricta*; 29A, 29B, *Navicula cuspidata var. heribaudii* (two different focal planes of same valve); 30, 31, *Navicula gallica*; 32, 33, *Nitzschia frustulum*; 34, *Nitzschia amphibia*; 35, *Nitzschia tropica*.

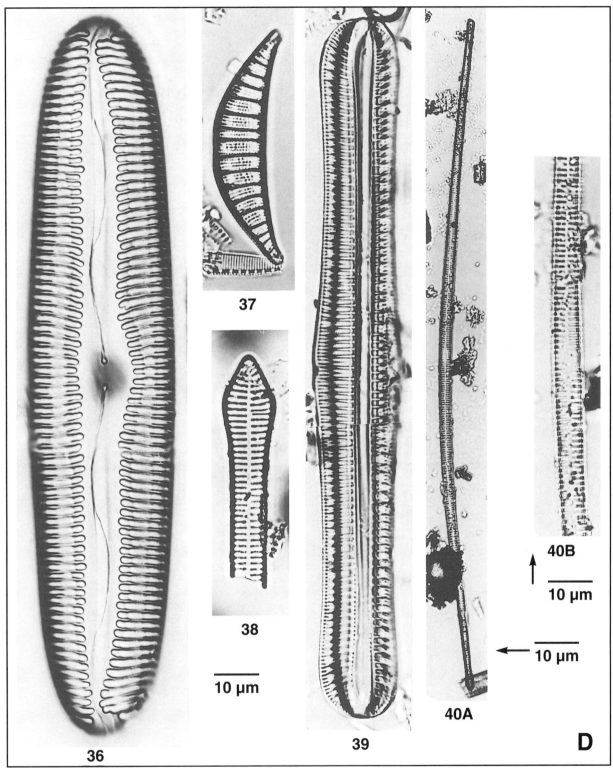

Figure 32D. 36, *Pinnularia viridis*; 37, *Rhopalodia gibberula*; 38, *Synedra capitata*; 39, *Rhopalodia gibba*; 40A, 40B, *Synedra ulna* (B is central area of same specimen as A).

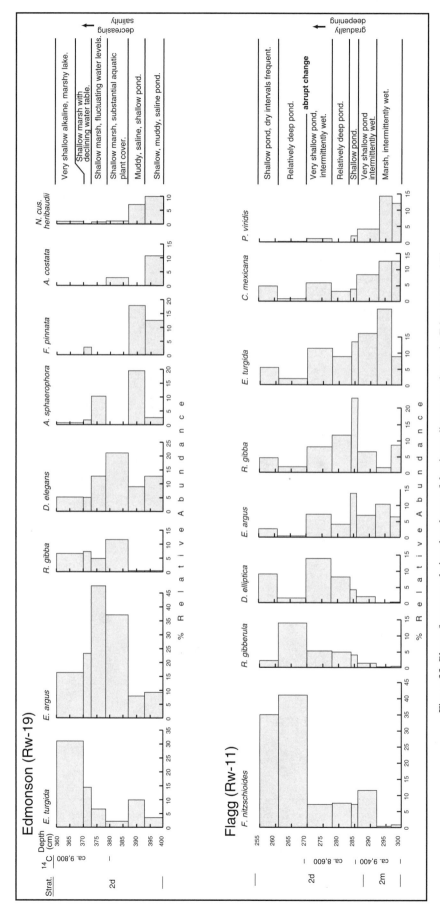

Figure 33. Plots of percent relative abundance of dominant diatom species at the Edmonson and Flagg sites in Running Water Draw (Fig. 4), showing long-term trends in species composition and paleoenvironments. "Deep" water is 1–3 m.

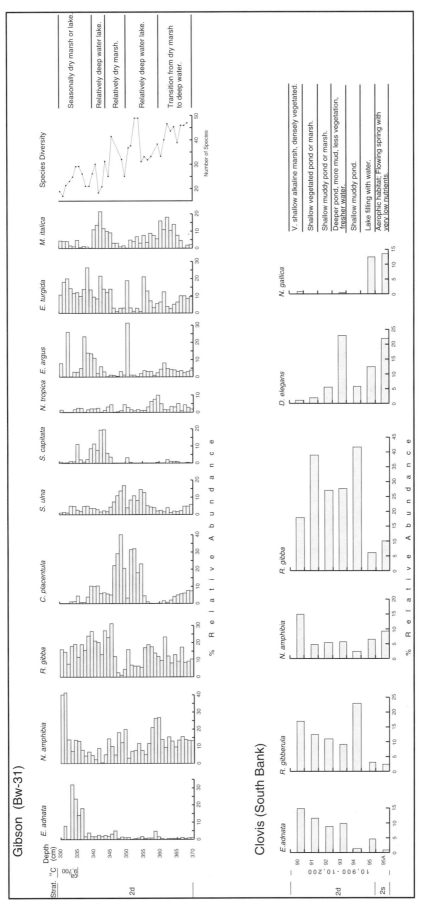

Figure 34. Plots of percent relative abundance of dominant diatom species at the Gibson and Clovis sites in Blackwater Draw (Fig. 4), showing long-term trends in species composition and paleoenvironments. The decline in species diversity through time is illustrated for the Gibson samples. "Deep" water is 1–3 m.

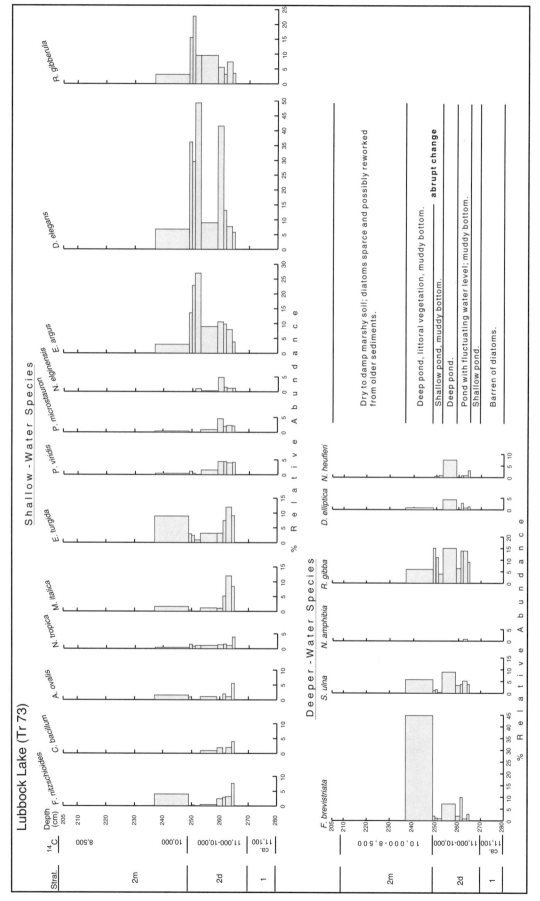

Figure 35. Plots of percent relative abundance of dominant diatom species at the Lubbock Lake site in Yellowhouse Draw (Fig. 4), showing long-term trends in species composition and paleoenvironments. "Deep" water is 1–3 m.

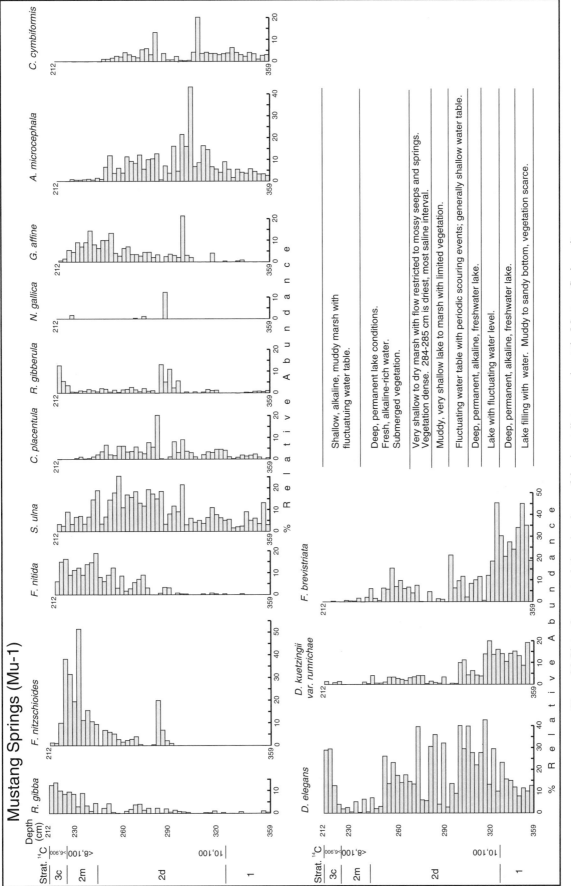

Figure 36. Plots of percent relative abundance of dominant diatom species at the Mustang Springs site in Mustang Draw (Fig. 4), showing long-term trends in species composition and paleoenvironments. "Deep" water is 1–3 m.

osmotic change. These characteristics indicate seasonal increases in carbonate and possibly chloride salt concentrations (alkalinity). An increase in alkalinity is tied to local ground-water changes that may be related to regional climatic trends.

The diatoms of all ten sites are remarkably similar in composition and suggest that there was a great similarity in the lacustrine habitats in the draws. There are wet and dry cycles that appear to be regional in character during the early Holocene (Table 34), and there is evidence of synchronous, regionwide deterioration of aquatic habitat associated with drying conditions between about 8,000 and 6,500 yr B.P.

SUMMARY, DISCUSSION, AND CONCLUSIONS

The principal goals of this study were to establish the stratigraphic record preserved in the draws of the Brazos and Colorado systems on the Southern High Plains and to use this record to reconstruct the late Quaternary environmental history of the region. The assembled data show that the stratigraphy along and between draws is remarkably similar in lithologic and pedologic characteristics, particularly considering the number and extent of draws studied (10 totalling >1,400 km in length) and the area involved ($\approx$40,000 km^2), but not all that surprising given the uniformity of the region's climate, hydrology, and surficial geology. These data further suggest that each draw underwent a similar, sequential, and dramatic evolution of the dominant depositional environments: in the late Pleistocene and early Holocene evolving from an alluvial environment into a complex lacustrine and palustrine environment, interfacing with an aggrading eolian environment in the early Holocene, and overwhelmed by the eolian environment in the middle Holocene. Hiatuses following the lake/marsh deposition and the eolian deposition are indicated by the development of soils. The changing depositional environments suggest other environmental changes such as shifts in regional vegetation and climate.

The changes in dominant depositional environments in the draws are very broadly synchronous, but there are distinct out-of-phase or time-transgressive relationships (Fig. 19) indicative of some regional variability in environmental evolution. These considerations raise three general questions: (1) What aspects of the stratigraphy are the result of regionwide environmental change? (2) What aspects of the stratigraphy are due to regionally variable or local environmental change? and (3) What aspects of the stratigraphy are due to other local factors such as damming, dune encroachment, or spring discharge? This final section of the monograph addresses these questions using all available data within the larger context of reconstructing the late Quaternary environmental history of the Southern High Plains. The discussion is organized around the dominant depositional environments and, therefore, there is some temporal overlap. Long-term late Cenozoic drainage evolution is summarized in the earlier section, "Geologic and Geomorphic Background". The following summary begins with the final stages of drainage evolution in the late Pleistocene and proceeds through the varying depositional envi-

ronments at the end of the Pleistocene and throughout the Holocene.

Late Pleistocene: Drainage evolution

Erosion and deposition by streams was the dominant geomorphic process affecting the draws in the latest Pleistocene, but the alluvial history of the draws is the least understood aspect of their late Quaternary record. In the late Pleistocene, before the last stage of incision of the draws, there was locally extensive deposition of lacustrine carbonate. These lake deposits show that water was impounded for a significant amount of time in paleobasins prior to final incision of the drainages. Data from a few mollusc assemblages (Clovis and Seminole-Rose sites; see previous section, "Paleontology, Paleobotany, and Stable Isotopes") indicate that the lakes had fluctuating levels and extensive wetlands along their margins. Uplands probably were well drained with no woody vegetation. Some of these paleolakes formed in older drainages (e.g., upper Blackwater Draw and upper Sulphur Draw). Data are not available to indicate whether all of the paleolakes were impounded along drainageways or if some of the basins were isolated and then integrated by the initial incision of each draw. The base of each lacustrine section (i.e., the floors of the respective paleobasins) is higher than the base of the local valleys fill, however, indicating that if the lakes were impounded along existing drainages then the paleovalleys were shallower than the present draws.

Radiocarbon ages from paleolake sediments incised by draws in three areas (Clovis and vicinity on upper Blackwater Draw; Lupton on lower Yellowhouse Draw; and Seminole-Rose on lower Seminole Draw) show that final downcutting along some reaches of the draws began during or after the period 20,000 to 15,000 yr B.P. The paleolake in upper Blackwater Draw in and near the Clovis site was in a paleovalley, but the geomorphic settings preceding downcutting at the other two sites are unknown.

A variety of scenarios can be offered to account for incision of the draws, including (1) internal, geomorphic controls; and (2) external, regional, environmental factors. The presence of incised lake deposits, in addition to previously noted lake basins along many of the draws (see earlier section, "Geologic and Geomorphic Background"), raises the possibility that incision was due to filling (by sediment and water) and overflow of the basins and subsequent breaching of sills. Alternatively, incision downstream, on the Rolling Plains, may have affected drainage development on the Southern High Plains. Caran and Baumgardner (1990, 1991), working in the Red River system of the Rolling Plains, believe most downcutting in the area occurred in the Holocene. Gustavson and Holliday (1988) argue that the postincision fill in the area is older than Holocene, however, and Blum et al. (1992) show that the Brazos River system immediately downstream from the draws underwent significant incision prior to ca. 13,000 yr B.P. Most streams on the Rolling Plains, in adjacent areas in Texas and Oklahoma, and in the northern Texas

TABLE 34. SUMMARY OF LATEST PLEISTOCENE AND EARLY HOLOCENE FLUCTUATIONS IN WATER LEVELS AND SOME LOCAL ENVIRONMENTS*

K (yr B.P.)	Running Water Draw			Blackwater Draw					Yh	Mustang Draw	
	Flagg	Edmonson	Plainview	Clovis	A.B. #1	Davis	Tolk	Gibson	Lubbock Lake	Mustang Springs	Wroe
8–					o saline marsh				x marshy (drying)	x marshy	
	X shallow										
	+ *deepest*										+ deeper (moist)
	x shallow								x marshy (drying)	x deep	
	x deepest / x lower		+ flowing								
9–	x deeper								x marshy (drying)	x marshy	+ deeper (wet)
	x deeper									x marshy to shallow	
	x deeper	o marsh or shallow pond									
		x deeper						x shallow		x fluctuating water level	+ shallow (dry)
		x shallow		x deeper				x deeper			
10–				x dry / x deeper			x marshy		x *deep* / x / x shallow / x	x deep / **x deep** / x deep / x shallow	+ v. shallow (v. dry)
				x shallow	+ shallow				x deep / x / x fluctuating / x	x muddy	+ deep (moist)
11–						x marshy			x marshy		

*Based on diatoms = x, molluscs = +, and ostracodes = o, grouped by draws, north to south, and arranged updraw to downdraw, from left to right (**bold** = ^{14}C age; nonbold = interpolated age based on ^{14}C chronology; *italic* = abrupt transition; marshy = wet but no open, standing water; shallow and deep refer to standing water; deep = >1 m). See Figure 4 for site locations.

Panhandle also underwent deep incision between 20,000 and 10,000 yr B.P. (Madole te al., 1991; Blum and Valastro, 1992; Mandel, 1992; Frederick, 1993b; Blum et al., 1994; Ferring, 1990, 1994; Humphrey and Ferring, 1994). Downcutting on the High Plains, therefore, may be due to a wave of incision (i.e., migration of nickpoints) that traveled up the Brazos and other drainage systems from the Rolling Plains to the High Plains in the latest Pleistocene. Base-level changes on the Rolling Plains also may have affected the draws prior to 20,000 yr B.P. Little data are available for the Brazos and Colorado River drainages of the Rolling Plains >20,000 yr B.P., but westward retreat of the Caprock escarpment and regional subsidence may have had a significant impact on the Red River drainage of the western Rolling Plains at some time between 300 and 40 ka (Caran and Baumgardner, 1990).

External environmental factors could have influenced incision as well. Sparse paleontological, paleobotanical, and pedological data for the period 20,000–12,000 yr B.P. indicate that the Southern High Plains probably was a grassland (perhaps a savanna or a sagebrush grassland) subjected to lower seasonal extremes relative to today, and, therefore, effective moisture was higher (Dalquest, 1986; Graham, 1987; Holliday, 1987, 1990b; Dalquest and Schultz, 1992; Hall and Valastro, 1995; Neck, 1987; and Neck in previous section, "Paleontology, Paleobotany, and Stable Isotopes"). Higher effective precipitation on a grassland likely produced higher runoff and discharge, but little increase in sediment yield. Such conditions in many semiarid settings result in channel incision (Knox, 1983; Bull, 1991), but increases in effective precipitation may not directly influence runoff and discharge in a significant way on the Southern High Plains due to the very small catchment areas of the draws. Springs fed by ground water, however, probably had a significant effect on stream discharge and the hydrology of the draws (discussed below), and ground-water levels and spring discharge are significantly affected by changes in precipitation in the region (Cronin, 1964; Fallin et al., 1987; Ashworth, 1991; Ashworth et al., 1991). Increased spring discharge would increase flow in the valleys without increasing sediment yield, likely resulting in incision.

The late Pleistocene downcutting of the draws was locally episodic, based on the presence of strath terraces along some reaches of the draws. Radiocarbon ages from fill in the bedrock channel at the Plainview site (Running Water Draw; Fig. 26B) show that most, but probably not all, local incision was complete by ca. 12,000 yr B.P.

Latest Pleistocene and early Holocene: Alluvial, lacustrine, and palustrine deposition

Alluviation (deposition of stratum 1) by a perennial, probably meandering stream, began locally by 12,000–11,000 yr B.P. Lenses of clay, often gleyed and sometimes organic-rich, near the top of the alluvium, along with diatoms and molluscs (see previous section, "Paleontology, Paleobotany, and Stable Isotopes"), document the intermittent development of marshes on an otherwise active flood plain. These marsh deposits within upper stratum 1 increase in thickness and number up section and denote the transition to the lacustrine and palustrine environments of strata 2 and 3. The end of alluviation was asynchronous along the draws, ceasing, for the most part, between 11,000 and 9,500 yr B.P.

Deposition of stratum 1 was followed by localized accumulation of diatomaceous muds and palustrine muds of stratum 2 or, more commonly, deposition of palustrine carbonates of stratum 3. Stratum 2 is found in isolated occurrences along the draws, mostly in the draws of the Brazos system, but also along lower Mustang Draw. The diatomaceous facies occurs in two different settings. At Clovis, Anderson Basin #1 and #2, the Gibson site, and Lubbock Lake, stratum 2 is found as a relatively extensive deposit extending across the bottom of the draw from valley wall to valley wall and along the length of the draw for at least 1–2 km. At the Lubbock Landfill and Mustang Springs stratum 2 is inset against stratum 1 (Figs. 20D, 27B, 30B), occupying relatively narrow reaches along the draw. At these sites 2d probably accumulated in paleochannels on top of stratum 1. The geometry of stratum 2m, the paludal facies, where it occurs without 2d, is poorly known because usually it was found in cores and not in exposures. The data suggest, however, that 2m accumulated in relatively confined areas on the floor of the draw.

The deposits of stratum 2 are indicative of fluctuating water levels based on geologic and pedologic characteristics (Holliday, 1985e), and based on evidence from diatoms, molluscs, and phytoliths (see previous section, "Paleontology, Paleobotany, and Stable Isotopes"). The diatomite and diatomaceous earth (2d) were deposited under standing water and the interbedded mud accumulated when water was at or below the surface. Extensive archaeological features in the mud lenses such as bone beds, provide further evidence that the muddy interbeds in 2d accumulated under subaerial conditions (Johnson and Holliday, 1980). The homogeneous, carbonaceous muds of stratum 2m, which overlie 2d or occur without it, represent slowly aggrading marshes, with water usually at or below the surface, but sometimes fluctuating between deep, standing water and subsurface water.

Water conditions fluctuated along with water levels. Periods of relatively deep, fresh water alternated with episodes of very low water and local desiccation (Table 34). Diatoms and molluscs (see previous section, "Paleontology, Paleobotany, and Stable Isotopes") indicate that water was sometimes saline during the low-water phases. Changes in water levels and water conditions probably were asynchronous between sites at a scale of centuries (Table 34), subject to influence by local factors such as the water table and geomorphology. Regional environmental trends are more apparent at the millennial scale (Table 34), however. At localities yielding diatom samples, the draw floors were marshy with shallow ponds >10,500 yr B.P. Water levels deepened and locally were at their deepest 10,500–9,500 yr B.P. Water levels were regionally variable after 9,500 yr B.P. Lubbock Lake and Mustang Springs, for example, had generally marshy conditions in the early Holocene, whereas the Flagg site had its deepest water at ca. 8,400 yr B.P. Marshy conditions with low or subsurface water eventually dominated the depositional environments of the early Holocene, even in areas that previously had fresh, standing water producing stratum 2d. The general history of stratum 2 deposition is therefore indicative of declining water levels and increasing water salinity through time. Stable-carbon isotopes from stratum 2 (see previous section, "Paleontology, Paleobotany, and Stable Isotopes") also show a shift from mesic plant communities that thrive on moisture to more arid-adapted species. Throughout deposition of stratum 2 the waters tended to remain neutral to slightly alkaline (see previous section, "Paleontology, Paleobotany, and Stable Isotopes").

There is no clear evidence for the mechanism impounding the waters that produced stratum 2. The Clovis, Anderson Basin, and Gibson areas are all within the Muleshoe Dunes, and evidence from Clovis and the Gibson site shows that some eolian activity was contemporaneous with the diatomaceous ponds. Dunes may have blocked the draw or, in the case of the Clovis site, the outlet channel. There is some evidence for such blockage by sand reworked from stratum 1 at Lubbock Lake (Holliday, 1985c). The gradient of most reaches probably was sufficiently low to impound water in shallow depressions, as described historically (Brune, 1981; GSA Data Repository 9541, Appendix D).

Stratum 2 accumulated in the last centuries of the Pleistocene and early millennia of the Holocene (Table 3). The stratum 1–2 transition dates to ca. 11,000 yr B.P. or greater throughout upper Blackwater Draw at and above the confluence with Progress Draw, at the Lubbock Landfill (on lower Blackwater), and at Lubbock Lake (on lower Yellowhouse). Otherwise, the stratum 1–2 transition occurred ca. 10,000 yr B.P. or later. Deposition of stratum 2 ended by ca. 9,000 yr B.P. at most sites, but continued until 8,500–8,000 yr B.P. at Clovis, Lubbock Lake, and Mustang Springs. Variations in the duration of the stratum 2 environments may be linked to the longevity of spring discharge, which is further discussed below.

Deposition of the marl facies of stratum 3 followed aggradation of either stratum 2 locally, or, more often, stratum 1. Palustrine carbonate was deposited along almost all reaches of all draws. Sedimentation then slowed or ceased, followed by formation of the weakly expressed Yellowhouse soil, characterized by a prominent low-carbonate, organic-rich A horizon. The A horizon formed under saturated conditions with the local water table immediately below the surface. Most of the marl deposition occurred between 10,000 and 7,500 yr B.P., but both the beginning and end of deposition is time transgressive along the draws (Table 3). The radiocarbon ages also show that deposition of strata 2 and 3 was partially contemporaneous.

The transition from flowing water (indicated by stratum 1) to standing water and marshy conditions (indicated by strata 2 or 3) represents a dramatic hydrologic change throughout the draws. The widespread decrease in the amount of water within the draws was related in part to a decrease in regional effective precipitation from the late Pleistocene to the early Holocene, based on paleontological data (E. Johnson, 1986, 1987c; Graham, 1987). Decrease in runoff, however, probably was not the sole or even dominant factor in changing the hydrology because, as noted, the catchment areas of the draws are small, essentially limited to the width of each valley. The hydrologic changes in the draws probably were the result of declining spring discharge. Ground-water levels and related spring activity respond directly and dramatically to significant changes in effective precipitation on the Southern High Plains (Cronin, 1964; Fallin et al., 1987; Ashworth, 1991; Ashworth et al., 1991). Reduction in effective precipitation reduces infiltration and ground-water recharge, which in turn reduces spring discharge. The local variability in the types and ages of the marsh and pond deposits probably is due to local environmental variability related to springs.

Diatoms and molluscs from stratum 2 (see previous section, "Paleontology, Paleobotany, and Stable Isotopes") show that the ponds and marshes in the draws were fed by springs, but the presence and importance of springs are documented as well by physical, locational, and temporal characteristics of the valley fill. Sites with organic-rich lake or marsh sediments generally are associated with spring activity. All sites with strata 2d or 2m are at or just below sites with historic springs (Table 35) or contain evidence of springs in the geologic record (Haynes and Agogino, 1966; Stafford, 1981; Meltzer, 1991). Most sites (19 out of 25) in the Texas portion of the Southern High Plains (where data on historic springs are available) with stratum 3m preserved also are associated with historic springs (Table 35). Accumulation of organic-rich sediment is a likely result of spring or seep discharge and presumably higher ground water, because such conditions would support more luxuriant plant growth and organic-matter production. Also, saturation of organic-rich sediment prevents oxidation and inhibits rapid decay of carbonaceous matter, allowing accumulation.

The relationship of lacustrine and palustrine sediments to spring activity also is suggested by regional variations in the lithology of the valley fill and by the number of historic springs in each draw. Lake and marsh sediments are more common in the draws of the Brazos drainage on the central Llano Estacado of Texas than in those of the Colorado drainage of the southern Llano (Fig. 37). Specifically, 26 of 44 sites (59%) in the Brazos drainage yielded lake or marsh sediments (Table 36). Historic springs also are more common in the Brazos systems (Fig. 17), where there is an average of 7 historic springs/100 km along the three draws, mostly in Running Water and Blackwater Draws (Fig. 17; Table 36). The distribution of presumed prehistoric springs is generally similar to that of historic springs, especially along middle Running Water Draw (compare Figs. 17 and 37). In the more intensively studied draws of the Colorado system (Sulphur, McKenzie, Seminole, and Lower Mustang), only 11 of 37 sites (30%) contained lake or marsh sediment (Fig. 37, Table 36) and there are only 4 historic springs/100 km. Historic springs are slightly less common in the other draws of the Colorado system, with 3 springs/100 km (Table 36). Moreover, most of the historic springs and most of the sites with lake or marsh sediment in the Colorado drainage are in Sulphur and Lower Mustang Draws.

TABLE 35. DRAW LOCALITIES WITH ORGANIC-RICH SEDIMENT AND/OR HISTORIC SPRINGS*

Draw Site[†]	2d	2m	3m	4m	5m	Historic Spring**
Blackwater						
Clovis	X	X				?
A. Basin #1	X	X				?
A. Basin #2	X					?
Baker			X			?
Davis		X				?
Birdwell					X	
Bailey Draw			X			X
Progress Draw		X	X			
Tolk		X			X	X
Halsell			X		X	X
Plant X					X	
Gibson	X	X			X	X
McNeese			X		X	X
Evans			X		X	X
Cannon		X				
LCU						X
Lubbock Landfill	X	X	X	?	X	X
BFI			X			X
Midland						
Boone			X			X
Mustang						
Ranger Hill			X			X
Glendenning			X			X
Mustang Springs	X	X	X		X	X
Walker			X			X
Upper Curtis Erwin					X	
Wroe		X	X			X
Lower Curtin Erwin					X	
Running Water						
Ned Houck			X			
Flagg	X	X				X
Sunnyside						X
Mandrell			X			X
Edmonson		X	X			X
Plains Paving		X			X	X
Quincy Street					X	X
Plainview		X			X	X
Plainview Landfill			X		X	X
Sulphur						
Bledsoe			X			X
Huckleby			X	X		
Sundown			X	X		
Brownfield			X			X
Lost Draw Flats						X
New Moore West						X
Yellowhouse						
Claunch		X	X		X	
Enochs			X	X	X	X
Brooks					X	
Anton			X			
Payne					X	
County Pit					X	
Lubbock Lake	X	X	X	X	X	X

*Figures 4, 17, 37.
†No study sites in McKenzie, Monument, or Monahans Draws yielded evidence for organic-rich sediment or historic springs; six historic springs are reported at other localities (not studied) among these four draws (see Appendix D).
§Strata deposited in marshes or ponds sustanied by ground-water discharge.
**Data not available for presence or absence of springs in New Mexico.

The geochronology of strata 2 and 3 also provide insights into the controls on their genesis. Radiocarbon ages along Running Water and Blackwater Draws, which have the best age control, indicate that the minimum age of the top of the basal alluvium (marking the end of alluviation) and the top of the marl (marking the end of marsh sedimentation) decrease downstream (Fig. 38). More limited data from Mustang Draw suggests a similar age relationship (Fig. 38). This phenomenon probably is linked to paleohydrological and paleoenvironmental changes. The regional direction of ground-water flow is from northwest to southeast (Gutentag et al., 1984) as is the dip of the bedrock aquifer system and the regional topographic slope. The surface effects of a regional decline in the water table (i.e., decreased spring and seep discharge) would be felt first to the northwest (up draw) and last to the southeast (down draw).

The lithostratigraphic and chronostratigraphic data, therefore, indicate a long-term, regional decline in the water table, generally moving in a time-transgressive pattern from northwest to southeast. This prolonged decline almost certainly resulted from a decrease in effective precipitation underway in the latest Pleistocene (ca. 11,000–10,000 yr B.P.), though the beginning date is unknown, and continuing into the early and middle Holocene. Regional effective drying and a declining water table resulted in alluviation terminating first updraw, with declining spring discharge no longer supporting a flowing stream, but only isolated ponds. As the water table continued to fall, the ponds declined and became marshes and eventually these dried as well, both effects felt first updraw. At about 9,000 yr B.P., for example, alluviation continued in middle and lower Running Water Draw as marl accumulated in the upper reaches. Haynes (1995) also proposed that a brief episode of desiccation and deflation occurred at Clovis, far up Blackwater Draw, ca. 10,000 yr B.P.

Diatom and mollusc data (see previous section, "Paleontology, Paleobotany, and Stable Isotopes") show that the ponds and marshes, particularly those that produced stratum 2, had broadly similar evolutionary histories with some local variations (Table 34). As alluviation waned marshy conditions emerged, followed by deepening water levels and eventual development of ponds. The timing of the onset and demise of the stratum 2 ponds and marshes was not synchronous and probably depended on the timing of the end of alluviation, and the nature and evolution of local geomorphic conditions that trapped the marsh and lake water.

The variation in the age of stratum 2 and its subunits is significant because previous investigators assigned regional climatic significance to this layer (Sellards and Evans, 1960; Wendorf, 1961a; Wendorf and Hester, 1975; Haynes, 1968, 1975). Patrick (1938) and Hohn and Hellerman (1961), for example, attributed general similarities of diatom assemblages and of changes in these assemblages through time to regional climatic shifts, rather than to local conditions. The diatom and mollusc data from the present study (see previous section, "Paleontology, Paleobotany, and Stable Isotopes") indicates some broad similarities in environments where stratum 2 accumulated, as described above and illustrated in Table 34. Microstratigraphic units and specific fluc-

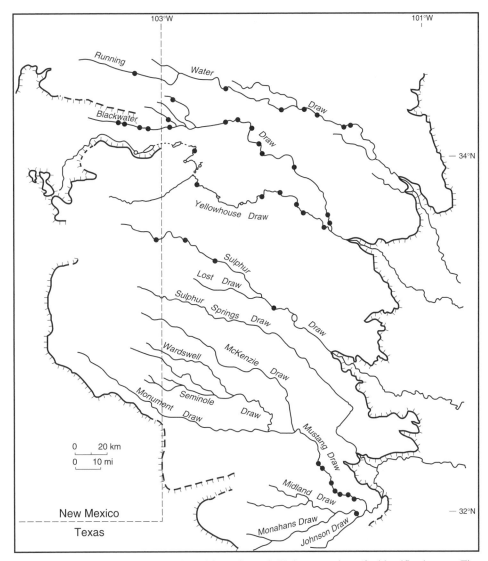

Figure 37. Locations of study sites with evidence for early Holocene springs (for identifications see Fig. 4 and Table 35). Compare with map of historic springs (Fig. 17).

tuations in water levels cannot be correlated, however, and likely represent local conditions.

The highly calcareous and very weakly carbonaceous nature of the marl (stratum 3c) contrasts markedly with the siliceous and relatively organic-rich character of stratum 2. Stratum 2 probably was deposited under cooler temperatures, which favor silica precipitation, and the marl was deposited as regional temperatures warmed, which would increase the carbonate concentration in water and decrease its solubility. Similar interpretations were proposed for Mustang Springs, based on diatoms (Winsborough, 1988). Molluscs from Plainview (stratum 2m; see previous section, "Paleontology, Paleobotany, and Stable Isotopes"), Anderson Basin #1 (stratum 2s; Drake, 1975), and Lubbock Lake (strata 2m and 3c; Pierce, 1987) also are indicative of increased temperatures, loss of soil moisture, and possibly drought across the strata 2–3 transition and during deposition of stratum 3c.

Very generally, the lithologic change from stratum 2 to stratum 3 is an indicator of climatic warming early in the Holocene, but if temperature or some other aspect of regional climate was the only control, then there should be a regional synchroneity in the geochronology of the two deposits. Strata 2 and 3 overlap in time, however, indicating that other factors are important. Besides lithology, the most obvious differences between the deposits is depositional environment and these differences could account for the temporal overlap. Where springs occurred and discharge remained vigorous, bringing abundant, cool water to the surface, soft-water ponds and marshes occurred. Where there was no spring effluence or when discharge waned, there were only seeps along the draws, the water warmed, and carbonates were precipitated. Diatoms and molluscs from stratum 3c (see previous section, "Paleontology, Paleobotany, and Stable Isotopes") indicate deposition under marshy, low-water conditions

TABLE 36. OCCURRENCE OF ORGANIC-RICH SEDIMENTS AND SPRINGS IN DRAWS OF THE TEXAS PORTION OF THE LLANO ESTACADO

Draw*	Length		Sites			OMSites/ 100 km	Springs	Historic Springs/ 100 km
	Total (km)	Tx Only (km)	Total	W/OM	%W/OM			
Rw	217	180	11	7	64	4	12	7
Bw	252	174	17	12	71	6	16	9
Yh†	154	128	16	7	44	5	4	3
Total		482	44	26	59		32	7
Su	175	108	9	4	44	4	9	8
Mk	144	135	8	0	0	0	2	1
Se	105	91	7	0	0	0	2	2
Mu	130	130	13	7	54	5	4	3
Total		464	37	11	30		17	4
Ss	224	224	--	--	--	--	7	3
Wd	74	63	--	--	--	--	2	
Mt§	77	77	2	0	0	1	1	1
Mn§	77	77	3	0	0	1	1	1
Md§	42	42	2	1	50	2	1	2
Total		420					12	3

*Draws listed north to south (see Fig. 3 for locations). Bw = Blackwater; Md = Midland; Mk = McKenzie; Mn = Monahans; Mt = Monument; Mu = Mustang; Rw = Running Water; Se = Seminole; Ss = Sulphur Springs; Su = Sulphur; Wd = Wardswell; Yh = Yellowhouse. Rw, Bw, Yh = Brazos system, all others = Colorado system. Within the Colorado: Mk, Se, Mt, Mu, Mn, Md = Mustang system.
†Data for Yh do not include the N. Fork.
§Mt, Mn, and Md draws were not systematically cored to an extent comparable to coring activities in the other draws, so the figures for numbers of sites with OM-rich sediments are not representative.

with occasional desiccation. Stable-carbon isotopes (see previous section, "Paleontology, Paleobotany, and Stable Isotopes") also are indicative of plant communities adapted to drier conditions. Variations in local depositional environments probably are related to the site-specific characteristics and history of groundwater changes, noted above, as well as local variation in water characteristics such as nutrient levels, pH, and depth.

To summarize, the latest Pleistocene and early Holocene were times of dramatic environmental change in the draws of the Southern High Plains and throughout the region. Between ca. 12,000 and 11,000 yr B.P. the draws incised to their deepest point. Competent graded streams were common on the floors of the drainages. Between 11,000 and 10,000 yr B.P., however, discharge began to wane, probably due to a decline in effective precipitation, which ultimately affected spring discharge along the valleys. The end of alluviation began first in the upper ends of the draws. Ponds and marshes then developed on the once-active stream beds.

The terminal Pleistocene mammal population was characterized by a Rancholabrean fauna until ca 11,000 yr B.P. when all the megafauna except *Bison antiquus* became extinct (E. Johnson, 1986). The late Pleistocene vertebrate faunas also are indicative of cooler environments in a savannah setting (Johnson, 1986, 1987a, c; Dalquest and Schultz, 1992). The one other significant, well-documented characteristic of the late Pleistocene mammal assemblage on the Southern High Plains is the arrival of Paleoin-

dian hunters and gatherers between 12,000 and 11,000 yr B.P. (Hofman et al., 1989).

Haynes (1991) proposes that the Southern High Plains was subjected to drought at the time of the Clovis archaeological occupation (11,300-10,900 yr B.P.; Haynes, 1993) and that megafauna extinctions at the end of the Pleistocene were linked to the drought. The data are from localities throughout the southwestern United States, including the Clovis site. Erosional disconformities in stratum 1 at Clovis are interpreted as deflation surfaces resulting from drought, and stratum 2d is interpreted as the result of increased spring discharge due to increased effective precipitation following the drought. Except for a layer of loess at the Miami site (Fig. 2; Sellards, 1938; Holliday et al., 1994), there are no other field data from the region to support an interpretation of drought during the Clovis period. Drought of sufficient intensity and duration to affect megafauna should produce a more obvious geologic record (e.g., eolian sediments and deflation surfaces) such as those resulting from the Holicene "Altithermal" drought (discussed below). An alternative interpretation for stratum 1 at Clovis is that it is an alluvial cut-and-fill sequence related to high, but fluctuating spring discharge. As effective precipitation declined, the basin ceased to discharge water and the channel below the basin became dammed with eolian sand. Ponds and marshes formed in the impounded basin and stratum 2d was deposited.

The environmental changes underway at the end of the

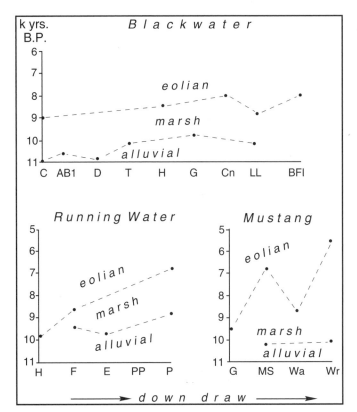

Figure 38. Diagram of radiocarbon means along Running Water, Blackwater, and Mustang Draws, showing the down-draw age variation of the top of the alluvium, top of the marsh deposits (marl), and base of the eolian fill. Sites (Fig. 4): Blackwater Draw, C = Clovis, AB1 = Anderson Basin 1, D = Davis, T = Tolk, H = Halsell, G = Gibson, Cn = Cannon, LL = Lubbock Landfill, BFI = BFI pits; Running Water Draw, H = Houck, F = Flagg, E = Edmonson, PP = Plains Paving, P = Plainview; Mustang Draw, G = Glendenning, MS = Mustang Springs, Wa = Walker, Wr = Wroe.

Pleistocene continued in the early Holocene and profoundly affected the landscape. Proxy data from the draws along with other paleontological data from the region (Johnson, 1986, 1987a; Neck, 1987) and climatic models (Kutzbach, 1987; COHMAP Members, 1988; Webb et al., 1993) indicate that temperatures were warming or were at least warmer in the summers and effective precipitation was declining. Vegetation on the uplands probably was short-grass prairie, based on phytoliths and molluscs (see previous section, "Paleontology, Paleobotany, and Stable Isotopes"), reflecting a continuation of the late Pleistocene prairie. Lack of opaque phytoliths, which indicate deciduous and coniferous woodlands, further suggests few trees on the landscape. Environments along the draws varied dramatically. Alluviation ended at all sites in the early Holocene. Some reaches had alternating ponds and marshes that eventually gave way to marshes, many of which were hard-water, alkaline marshes, and some were saline. Occasionally some reaches of the valleys were desiccated. By the end of the early Holocene hard-water, alkaline marshes and localized desiccation typified the draws.

Early to middle Holocene: Eolian deposition

Wind-derived sediment is the most common deposit found in the draws. Eolian deposits are a component of all five lithostratigraphic units, increasing in significance from the end of the Pleistocene (stratum 2s) into the middle Holocene (stratum 4s; Fig. 19). Evidence for eolian activity is least common in stratum 1, but eolian sediments became an important component of the valley fill early in the Holocene. Eolian facies of strata 2 and 3 are well documented, but not widespread or common, and eolian sediment probably is a significant constituent of the carbonate facies of stratum 3. Eolian sedimentation began by 10,000 yr B.P. at Clovis, 9,900 yr B.P. at Gibson, 9,000 yr B.P. at Lubbock Lake, 8,800 yr B.P. at Lubbock Landfill, and after 8,800 yr B.P. but well before 6,800 yr B.P. at Plainview (Table 3).

Soils in the sandy, eolian facies of strata 2 and 3 are indicative of relatively dry, well-drained conditions along valley-margins which co-occurred with ponds and marshes along the valley axes. As the valley floors dried and eolian sedimentation increased, these drier valley-margin microenvironments probably expanded. Dunes may have been common along the valley margins and soils would have been excessively drained. Similar semiarid environmental conditions were found in the draws earlier in this century (Fig. 18).

Stratum 4 is the principal eolian unit in the valley fill, composed mostly of wind-derived fines. Stratum 4 locally includes early Holocene sediments not obviously related to valley axes facies of strata 2 or 3, but largely represents the culmination of Holocene eolian activity. Stratum 4 occurs as sheets draped across the valley floor or as dunes where the draws traverse dune fields (upper and middle Blackwater Draw, upper Sulphur Draw, and middle Monahans Draw). In upper and middle Blackwater Draw, however, dunes or sheet deposits of stratum 4 (and 5) are very localized and stratum 3 is exposed along most of this reach of the draw.

The deflation of the High Plains surface that produced the eolian sediment probably resulted from a reduction in vegetative cover and loss of soil moisture, which allowed winds to scour the landscape. The vegetation change was most likely in response to increased temperatures and lower effective precipitation relative to the preceding or following periods. Studies of native grasslands on the Great Plains clearly show that the vegetative cover is significantly reduced within several years of the onset of drought, allowing wind, a common co-occurrence of drought in the region, to deflate the surface and produce significant quantities of eolian sediment (Weaver and Albertson, 1943; Tomanek and Hulett, 1970; Holliday 1989a). Even on the floors of the draws the vegetation apparently changed; carbon isotopes from Lubbock Landfill, Lubbock Lake, and Mustang Springs indicate a marked shift toward warm-season grasses early in the Holocene (ca. 10,000–7,500 yr B.P.; see previous section, "Paleontology, Paleobotany, and Stable Isotopes"). Eolian sedimentation in the draws was widespread by 7,500 yr B.P., but the most intense deflation, judging from the extent of eolian deposits, was

6,500–4,500 yr B.P. This indicates that deflation was due more to a reduction in plant cover rather than the shift to more warm-season grasses because that event occurred earlier. The vegetation change probably contributed to the initial, early Holocene eolian sedimentation, however. The amounts of middle Holocene eolian sediment preserved in the draws and dunes of the Southern High Plains indicate that wind erosion of the High Plains surface probably was quite severe. This severity can be gauged relative to the very minor geologic record of the Dust Bowl of the 1930s.

There was little moisture on the floors of the draws in the late middle Holocene (6,000–4,500 yr B.P.), except for the marsh muds at Lubbock Lake (stratum 4m) and Wroe (stratum 3m), and possibly Plainview. These deposits, with some interbedded alluvial sand and aquatic molluscs at Lubbock Lake, provide the only evidence in any of the draws of continued spring discharge through the prolonged arid stage. This is testimony to unusually strong and persistent spring discharge at these sites.

Beyond eolian sedimentation and declining water levels, there is other evidence of the impact of drought and dustiness in the draws. Teeth from modern bison in stratum 4 at Lubbock Lake show considerable wear, attributed to the high amounts of airborne grit in forage (Johnson and Holliday, 1986). Perhaps the most telling evidence of the impact of drought is in the form of hand-dug wells found on the floors of the draws at the Clovis site (Evans, 1951; Green, 1962b), Mustang Springs site (Meltzer and Collins, 1987; Meltzer, 1991), and 41MT21 in lower Sulphur Springs Draw (Quigg et al., 1994), and perhaps elsewhere (Honea, 1980; Meltzer, 1991). At least 19 wells were found at Clovis, excavated some time after ca. 8,000 yr B.P. (possibly after ca. 7,000 yr B.P.), but before ca. 5,000 yr B.P. (Haynes, 1995). Nearly 60 wells were exposed at Mustang Springs, dated to between ca. 6,800 and ca. 6,600 yr B.P. (Meltzer, 1991). Two wells were identified at 41MT21, one dating to the early Holocene and the other to the middle Holocene (Quigg et al., 1994). The wells at all of these sites are believed to be the result of excavation by aboriginal occupants trying to reach a water table lowered by prolonged drought (Meltzer, 1995).

The data from the draws clearly indicates that the early and especially the middle Holocene were times of maximum late Quaternary aridity. Climate models for the interior of North America at 6,000 yr B.P., relative to 9,000 and 3,000 yr B.P., indicate that seasonality was more pronounced than at present, resulting in a net annual moisture deficity greater than that of today (Kutzbach, 1987; COHMAP Members, 1988; Webb et al., 1993).

The middle Holocene environment of the Southern High Plains roughly corresponds to the classical definition of the "Altithermal" (Antevs, 1955), although soils within stratum 3 are indicative of periods of very slow aggradation or nondeposition, showing that eolian sedimentation and presumably drought were episodic. The geochronology and areal extent of these episodes is very poorly known, but proposals of a regionwide "two-drought Altithermal" (Holliday, 1985c; Johnson and Holliday, 1986) no longer are tenable. The chronostratigraphic data suggest that droughts varied in intensity and probably in geographic extent though time.

Late Holocene: Stability and alluvial, eolian, and palustrine deposition

By ca. 4,500 yr B.P. eolian sedimentation ceased or slowed and pedogenesis began in stratum 4. These events indicate that the density of the short-grass prairie increased sufficiently to inhibit wind erosion, stabilize the draws, and promote soil formation. Essentially, the environment of most of the late Holocene was similar to today's environment.

Pedogenic modification of stratum 4 was considerable. Pedogenic features dominate the field characteristics of stratum 4 and include iron oxidation, subangular blocky and prismatic structure, significant accumulation of organic matter in the A horizon, and significant translocation and illuviation of both clay and calcium carbonate. Mollic, calcic, and argillic horizons are common but not ubiquitous. All of these characteristics suggest pedogenesis under well-drained conditions.

Considerable variability in pedogenic features is apparent among the study sites. In general, there is weaker expression of pedologic characteristics on slopes and very strong expression along valley axes. These variations likely reflect differences in parent material and slope position. The valley-axis facies of stratum 4 typically is finer than the valley-margin facies. This results in valley-axis soils with more strongly expressed structure and darker A horizons. Slope position probably is a more important factor, however. Surface horizons along the valley margins are thinner owing to erosion and a drier soil-moisture regime. Organic matter accumulated rapidly and degraded more slowly along the valley axis, which is relatively more moist and is at the foot of the eroding slopes. Through-flow of water from soils on the slope to those on the valley floor also probably contributed to enhanced clay and carbonate accumulation in the valley-axes soils.

Pedogenesis was the dominant geomorphic process in the draws during the late Holocene, but erosion and deposition locally were significant. Channel cutting is notable at Houck and the Quincy Street section on Running Water Draw, at Lubbock Lake on lower Yellowhouse, along lower Sulphur Springs Draw (leaving the younger valley fill as a terrace; Frederick, 1994), and at Mustang Springs on lower Mustang Draw. The erosion on the lower reaches may be due to headward erosion that began in the deep canyons to the southeast, where the drainages cut the High Plains escarpment. Field evidence at Lubbock Lake and Mustang Springs suggests, however, that the erosion may be due to local head cutting induced by spring discharge. Large depressions cut into older valley fill mark the now-dry historic springs at the sites (Figs. 14E, 30A, B). Historic springs also were present near the Quincy section. Data on historic springs in New Mexico are not available, however, and such relationships at Houck can not be established.

Muddy facies of stratum 5 (5m) commonly occur at or below historic spring sites (Table 35) and the muds probably were deposited in spring-fed marshes. At Lubbock Lake the muds were deposited throughout much of the late Holocene (beginning at least 3,900 yr B.P.). Stratum 5m accumulated for the past 3,500 years at Plainview, between 3,000 and 500 yr B.P. on lower Sulphur Springs Draw (Frederick, 1994), and for the past 2,000 years at Mustang Springs. Furthermore, 5m in lower Yellowhouse and lower Sulphur Springs Draws locally is inset against the valley fill, and the data from Lubbock Lake and Mustang Springs show that there were multiple episodes of erosion and deposition during accumulation of the muds.

Some of the cutting and filling in the draws probably is related to local resurgence of spring activity. Increased spring discharge was due to increased effective precipitation and infiltration in the late Holocene which raised water tables. Multiple cuts and fills may be a result of fluctuating spring discharge, which, in turn, may be related to climate changes. Data are insufficient to pursue this relationship, however, beyond several points discussed below.

In most settings where the stratum 5 muds occur, they were deposited slowly on top of stratum 4, keeping pace with genesis of the Lubbock Lake soil. The result is that stratum 5m is now pedogenically welded to stratum 4, i.e., the muds form the overthickened A horizon of the Lubbock Lake soil.

The eolian facies of stratum 5, like the older eolian deposits, may represent episodes of aridity. At the Lubbock Landfill and at Lubbock Lake, these deposits are underlain by sandy and gravelly alluvial sediments that are slope-wash facies of stratum 5 and are suggestive of destabilization of the landscape surrounding the draws. Such destabilization may represent the early stages of drought. If so, the geomorphic effects of these dry periods were very localized. Dating indicates that eolian deposits accumulated, perhaps in several episodes, between 3,000 and 2,000 yr B.P. (Plainview, Lubbock Landfill, and Evans), after 2,000 yr B.P. (Plainview), and again after 1,000 yr B.P., but probably before 500 yr B.P. (Flagg, Evans, Lubbock Lake, Mustang Springs, possibly Clovis). There is weak isotopic evidence for drying at Lubbock Lake between 2,000 and 500 yr B.P., but the best evidence for regional aridity at the site in the late Holocene is for the period 1,000–500 yr B.P.

Return to increased effective moisture around 500 yr B.P., at least locally, is indicated by the carbon-isotopic data from Lubbock Lake as well as evidence for expansion of the marshes. Deposition of stratum 5m at Lubbock Lake became much more extensive after 500 yr B.P. The only other deposits that provide evidence of regional drying in the late Holocene are sands in the dune fields of the region, but these strata have poor age control.

Investigators working in the Texas Panhandle and on the Rolling Plains east of the Llano Estacado, report late Holocene stratigraphies and climate reconstructions partially in agreement with and partially in contrast to those reconstructed from the Younger Valley Fill. Baerreis and Bryson (1965, 1966) proposed that the Canadian River valley in the Texas Panhandle received greater precipitation (relative to immediately preceding and following periods) between 700 and 500 yr B.P., based on radio-

carbon dating and archaeological interpretations. No proxy indicators were offered, however. North of the Canadian Valley, in Palo Duro Creek, Frederick (1993b) reports late Holocene (<2,000 yr B.P.) colluvial and eolian sediments, presumably indicating departures toward aridity and landscape stability, but a more precise geochronology is not available. The prairie vole *Microtus ochrogaster* appears in archaeological contexts in the Panhandle between 2,000 and 1,000 yr B.P., considered an indicator of conditions more moist than today, then disappears, suggesting a drying trend (Willey and Hughes, 1978; Hall, 1982; Johnson, 1987a). Hall (1982) also presents evidence from analyses of pollen and landsnails at sites in northeastern Oklahoma for a similar scenario of climatic trends over the past 2,000 years. In western Oklahoma, Hall and Lintz (1984) report evidence from radiocarbon-dated buried trees, buried soils, pedogenic carbonate accumulation, and molluscs for drying between 3,200 and 2,600 yr B.P., a more moist climate (relative to today) 2,000–1,000 yr B.P., and drying after 1,000 yr B.P. In contrast, Ferring (1990) and Humphrey and Ferring (1994), using isotopic, geologic, faunal, and pollen data, from a site in north-central Texas, reconstruct a relatively more moist late Holocene 4,000–2,000 yr B.P., drying 2,000–1,000 yr B.P., and a return to more moist conditions in the past 1,000 years.

Reconciling the conflicting environmental reconstructions for the late Holocene is not possible with the present database. The differences may be partially due to comparisons of different sets of data (isotopes, sediments, pollen, molluscs, and soils) and comparisons of sites in different local settings (valleys without perennial streams, valleys with perennial streams, and rockshelters) and different regional settings (the High Plains and the Rolling Plains) subject to influence by different air masses. The localized and nonsynchronous nature of late Holocene erosion episodes also may be linked to high magnitude and low frequency (and therefore localized) storms. The advent of such storms in the continental interior appears to be largely a late Holocene phenomenon (Knox, 1983). Heavy runoff from relatively well-vegetated uplands probably had low sediment loads and therefore were highly erosive as they flowed down the draws. The localized, discontinuous, and asynchronous stratigraphic record of late Holocene valley fill in itself may be a clue to the reason various paleoenvironmental records are seemingly at odds with one another: paleoenvironments and environmental changes in the late Holocene were localized and relatively minor compared, for example, to the widespread and more substantial changes of the late Pleistocene and early Holocene.

Conclusions

The draws of the Southern High Plains have a remarkably complete late Quaternary stratigraphic record. The valleys contain sediments and soils that are spatially continuous for scores to hundreds of kilometers and together provide a temporally continuous geologic record of the past 12,000 years. The strata appear to be unique to the draws of the Brazos and Colorado

River systems. Below the Caprock escarpment these drainage systems have cut-and-fill sequences typical of alluvial systems (Madole et al., 1991; Blum and Valastro, 1992; Blum et al., 1992; Mandel, 1992). The draws of the Red River system, on the northern Llano Estacado, also appear to be dominated by alluvial sequences, though data are very limited (Finley and Gustavson, 1983). Data from drainages that flow west from the Llano Estacado into the Pecos River, also are indicative of alluvial stratigraphic sequences markedly different from the draws described in this paper (Smith et al., 1966; Parry and Speth, 1984).

The varied deposits and soils, along with paleobotanical, paleontological, and archaeological remains provide abundant clues to the paleoenvironmental history of the draws and the neighboring uplands of the High Plains surface. Indeed, the draws contain the most complete and sensitive regional environmental record so far available. These data provide the bases for addressing some specific, significant research questions regarding regional versus local environmental changes, hydrologic changes, the significance of eolian sediments, and paleoenvironmental reconstructions of earlier workers.

The lithostratigraphic and pedostratigraphic records in the draws are remarkably similar throughout the region, but there is some significant chronostratigraphic variation along and between the draws. The chronostratigraphic variability along the draws is a clue to deciphering the significance of the abrupt shift from flowing to standing water at the end of the Pleistocene and in the earliest Holocene. Ultimately, this hydrologic change was brought about by a climate change to reduced effective precipitation, but the immediate cause was reduced spring discharge. With regional drying recharge was unable to keep pace with discharge, resulting in a decline in the zone of saturation. This decline lead to declining spring discharge. Variation in the timing of this decline was responsible for the variation in the age of the resulting deposits. A regional drying trend, in the form of increased seasonality with overall warming and decreased effective precipitation, clearly was underway throughout this time, but local environmental factors played a key role in controlling depositional environments and the resulting stratigraphic record at any one time and place.

Regional environmental factors played an increasingly important and direct role in creating the stratigraphic record of the draws in the early and middle Holocene. The draws dried out as the climate warmed and dried. This climatic trend also resulted in wind erosion of the High Plains surface and deposition of eolian sediments in the draws. Eolian sediments mixed with pond and marsh deposits in segments of draws where spring discharge continued even as deflation affected the uplands. Through the early Holocene, as ponds and marshes dried, eolian sediments became more common until they completely dominated the middle Holocene. This combination of declining spring activity and increased eolian sedimentation testifies to the significance and severity of middle Holocene drought.

Both regional and local environmental conditions are well recorded in the late Holocene record of the draws. By ca. 4,500 yr B.P. eolian sedimentation in the draws ceased as effective precipitation increased. The draws became stable geomorphic settings with little erosion or deposition. Pedogenesis throughout the late Holocene strongly modified the middle Holocene deposits. The increased regional moisture lead to resurgence of spring activity in the draws, an event documented in the geologic record by marsh and pond muds and, in the historic record, by the many accounts of springs and lakes. Brief and probably localized climatic departures towards aridity occurred episodically in the late Holocene, leaving relatively minor and localized depositional traces.

The variation through time in the significance of local versus regional environmental impacts in the draws, along with the asynchronous characteristics of the stratigraphy, will be very important considerations in any future research in the draws. These considerations will be of particular significance to archaeological investigations for several reasons. Such investigations traditionally rely on stratigraphic correlation for dating (e.g., Sellards and Evans, 1960; Haynes, 1968; Johnson, 1974), but the boundaries between stratigraphic units in the draws can vary in age by as much as 1,000 to 2,000 years. Sedimentology, pedology, and stratigraphy have long been used for environmental reconstructions in the region (e.g., Stock and Bode, 1936; Sellards and Evans, 1960; Haynes, 1975; Holliday, 1985c), but investigators will have to be aware that certain variations may reflect more local environmental changes or events, and are not necessarily indicative of regional environment and climate.

Springs also appear to be keys to archaeological site location. Sites in draws with evidence for repeated occupation (Clovis, Gibson/Marks Beach, Lubbock Lake, and Mustang Springs) also have evidence of historic and prehistoric springs. This is not surprising given the importance of water to human survival, but indicates that evidence for ancient springs is more important for prediction of archaeological site location than is topographic position in or along the draws (cf. Stafford, 1981).

Ancient and modern springs were not evenly distributed along the draws, but concentrated along reaches of Running Water, Blackwater, and Lower Mustang Draws (Figs. 17, 37). The reasons for this pattern probably are related to the sedimentology, stratigraphy, and hydrology of the regional aquifer as it interacted with the draws, the specifics of which are unknown. One controlling mechanism is suggested, however, by site-specific stratigraphic data from the draws: there is no Ogallala Caprock calcrete at the three most intensively studied sites of ancient and modern spring flow (Clovis, Lubbock Lake, and Mustang Springs). Absence of this dense rock may be one of several key variables that promoted spring discharge in the draws (Meltzer, 1991).

The stratigraphic and paleoenvironmental schemes proposed in this volume broadly follow the pioneering reconstructions of the paleoecology of the Llano Estacado (e.g., Wendorf, 1961b, 1975b; Wendorf and Hester, 1975), but vary considerably in detail. The various late Pleistocene and early Holocene climatic intervals proposed in those early efforts (Table 2) are not identifiable in the composite stratigraphic record that is emerging. This discrepancy may be due to the inherent lack of resolution in the lithostratigraphic record, but more likely it reflects problems of dating, stratigraphic correlation, and pollen preservation in the work that led to the earlier schemes (Holliday, 1987; Hall and Valastro, 1995; see previous section, "Paleontology, Paleobotany, and Stable Isotopes"), and the great deal of research done since then.

The data presented in this study and in Hall and Valastro (1995) provide additional evidence that the Llano Estacado has been some type of grassland for at least the past 20,000 years, although the specific floristic composition undoubtedly varied over this time. These data directly counter the interpretation of a cold, wet boreal forest on the Llano Estacado at the close of the Pleistocene (e.g., Wendorf, 1970). The interpretation offered here more closely approximates that of Wendorf and Hester (1962, p. 159; and Antevs, 1954, before them) who proposed that the terminal Pleistocene environment "consisted of a savanna grassland with abundant ponds and streams" and trees along the escarpments to the north and west.

Additional stratigraphic and paleoenvironmental research on the Southern High Plains certainly is warranted and should shed additional light on the paleoenvironmental history of the region. The fill in the draws historically is linked to the human prehistory of the area as a result of the discovery of several renowned archaeological sites in these settings. Continued investigation of the rich and long archaeological record requires a better understanding of the environmental history at site and regional scales. Intensive, site-specific investigations similar to those at Clovis and Lubbock Lake should prove to be particularly informative.

The draw study also provides insight into the suitability of various paleoenvironmental indicators. Vertebrate, diatom, mollusc, and ostracode remains are well preserved locally and have proven very useful in reconstructing past environments. Recovery of adequate vertebrate samples requires processing large amounts of sediment, however, and the diatoms, molluscs, and ostracodes provide generally local environmental data. Some of these remains may be able to provide further information on ancient water chemistry, temperature, and depth, however. The diatoms, molluscs, and ostracodes also are sensitive to microhabitat variability, but in this study were not sampled to detect such variability. Future research could focus, for example, on microenvironments from the valley axes to the valley margins or within and among the ancient ponds and marshes that once existed in the draws. Recovery of suitable pollen and insect samples has proven frustrating and disappointing. The utility of these environmental indicators on the Southern High Plains probably will be rare and site specific. Phytolith and stable-carbon isotope analyses, however, hold considerable promise as methods widely applicable for obtaining clues to regional paleoenvironments.

The research results discussed in this volume also will provide the basis for work in the other loci of late Quaternary deposition: the dunes of the uplands and the lake sediments in the playas. Relative to the draws, considerably less is known about the history of the dunes and playas, although they are more widespread and ubiquitous geomorphic features than are the draws. The dunes should yield additional data on the regional response to drought and wind erosion. The playas, because they are sites of essentially continuous sedimentation in a more or less unchanging depositional environment, should provide a good contrast to the record preserved in the draws. Preliminary test results also suggest that the playas may yield good phytolith and stable-carbon isotope records.

Understanding the response of the landscape to environmental change takes on considerable importance in light of growing concerns over the future environment of the Plains and the response of the landscape to drastic climate changes (Borchert, 1971; Bark, 1978; Manabe and Wetherald, 1986; Hansen et al., 1988; Schneider, 1989; Smith and Tirpak, 1990). Future studies of the dunes and playas probably hold the keys to further understanding of the history and climatic sensitivity of the High Plains landscape.

APPENDIX 1. FIELD METHODS AND DESCRIPTIONS OF CORES AND SECTIONS

Vance T. Holliday and Peter M. Jacobs

Stratigraphic, sedimentologic, and pedologic data were collected from 407 cores and exposures at 110 sites, including localities investigated prior to 1988. Coring sites were approximately evenly spaced along each draw in a manner to provide an adequate sample, but also to allow prompt completion of the project. The spacing of the coring sites also was determined somewhat by the presence of artificial exposures; limited or no coring was done when a large pit was available. Otherwise, two to five cores usually were taken in a transect across the draw at each site. The number of cores depended on the width of the draw and data available from nearby sites. At several sites that were of particular interest or topographic complexity, more than five cores were taken over an area larger than a single transect. Each coring site was given a name, based on that of either the landowner, a nearby town, or some other landmark. Cores were numbered consecutively for each draw and each draw was given a two-letter abbreviation. For example, a coring site in Running Water Draw, near the town of Bovina, Texas, was named the Bovina site and cores Rw-5, Rw-6, and Rw-7 were taken at that site. Because the cores were numbered in the order in which they were taken, and because some sites were cored on several occasions, some sites have cores that are not in numerical order. For example, the Gibson site in Blackwater Draw covers a 5 km reach of the draw and was cored or trenched during all five field seasons. The site has numbers Bw-19, 20, 21, 22, 31, 49, 56, 57, 61-69, and 76-90.

Backhoe trenches, individual sections at some quarries, and bucket-auger sites also were numbered as part of the same identification scheme. At Gibson, for example, Bw-56 and Bw-61 are bucket-auger sites and Bw-78 is a trench. At the Quincy Street site in Plainview, Rw-24 through Rw-30 are backhoe trenches. Sections at a few sites with large exposures (Plainview, Clovis, Lubbock Landfill, and Lubbock Lake) were identified using different schemes and these schemes are discussed in the individual site descriptions (GSA Data Repository 9541, Appendix A).

All cores and sections were measured and minimum basic descriptive data (i.e., pedologic horizonation and structure, boundaries, bedding, and color characteristics) were recorded in the field. Samples from selected cores and sections were also brought back to the laboratory for further description and characterization using standard pedologic and geologic nomenclature (AGI, 1982; Birkeland, 1984; Soil Survey Staff, 1990; Soil Survey Division Staff, 1993) with two modifications regarding buried soils and stages of carbonate accumulation. Buried soils are numbered with a suffix following the "b" according to their stratigraphic position below the surface (e.g., A-Bw-Ab1-Btb1-Cb1-Ab2-Btb2).

Stages of calcium carbonate accumulation are described following the system of Gile et al. (1966) as modified by Machette (1985) and Shroba (in Birkeland, 1984, table A-4). The schemes of Gile et al. and Machette are designed for use in soils formed in parent materials generally coarser than those found on the Southern High Plains and include the development of carbonate nodules. Nodules are rare in even relatively strongly developed calcic horizons in comparatively fine-grained sediments, such as in calcic horizons with as much as 25% calcium carbonate and 50% carbonate films and threads (visual estimate). A modification of the existing systems of carbonate accumulation morphology is presented for fine-grained (medium sands and finer), nongravelly sediments (from Holliday, 1982):

Stage If (filamentous) <15% (by volume, visual estimate) filamentous threads, films, and/or coatings of $CaCO_3$ on and/or in peds; <10% $CaCO_3$ in whole sample; values (dry color, or whole soil crushed) are <6.

Stage IIf (filamentous) At least two of the following characteristics: between 15–50% (by volume, visual estimate) filamentous threads, films, and/or coatings of $CaCO_3$ on and/or in peds; 10–25% $CaCO_3$ in whole sample; values (dry color, of whole soil crushed) are 6–9.

Soils with intermediate characteristics are designated Stage If+. Soils with calcic horizons more strongly developed than Stage IIf, but without carbonate nodules, can be designated as Stage IIf+ or Stage IIIf if carbonate essentially coats the entire horizon.

Included in this appendix are descriptions of cores and sections with some age control or which are representative of undated draws (Tables A1.1–A1.33). Descriptions of all 110 study sites are presented in Appendix A (GSA Data Repository 9541).

Notes

(1) *Data on boundary topography were not recoverable from examination of cores. Therefore, boundary characteristics for cores include only distinctness.*

(2) *All horizons are calcareous unless otherwise noted.*

(3) *Abbreviations for all descriptions:*

Color (Munsell): m = matrix; c = carbonate; + = slightly redder than indicated hue.

Text. (Texture): v = very; f = fine; m = medium; S = sand; C = clay; L = loam; SC = sandy clay; SCL = sandy clay loam; SL = sandy loam; LS = loamy sand; CL = clay loam; SiC = silty clay; SiL = silty loam; SiCL = silty clay loam; + = high clay content within given textural category.

Str (Structure): Grade—1 = weak; 2 = moderate; 3 = strong. Class—f = fine; m = medium; c = coarse. Type—sg = single grain; gr = granular; pl = platy; sb = subangular blocky; ab = angular blocky; pr = prismatic; m = massive.
More than one designation for structure indicates compound structure.

Cons. (Consistence): Dry—lo = loose; so = soft; sh = slightly hard; h = hard; vh = very hard. Moist—fr = friable; fi = firm.

Bndy (Boundary): distinctness—a = abrupt; c = clear; g = gradual; d = diffuse; topography—v = very; s = smooth; w = wavy; i = irregular.

Remarks: com = common; concr = concretions; cont = continuous; carb = carbonate; dia = diameter; disc = discontinuous; irreg = irregular; max = maximum; OM = organic matter; pri = primary.

TABLE A1.1. DESCRIPTIONS OF ANDERSON BASIN #1, SECTION BW-71, AND ANDERSON BASIN #2, CORE BW-58

Bw-71

Strat.	Horizon	Depth (cm)	Color Dry Moist	Text.	Str.	Cons.	Bndy.	Remarks
3c	C	0-18	10YR 8/2 7/2	LS	m	fr	cw	Common limonite stains @ 6-13cm.
2s or 3s	Agb1	18-28	10YR 6/2 4/2	LS	1msb	fr	ci	Common limonite stains; C14 sample.
	C1b1	28-50	10YR 7/2 6/2	S	sg	lo	as	Common, discont. limonite stains 28-32cm; faint bedding ave. 2-5mm thick.
	C2b1	50-52	10YR 6/2 5/2	SL	m	fr	cs	V. faint bedding.
	C3b1	52-87	10YR 7/2 6/2	S	sg	lo	ai	Few, discont. clayey, muddy laminae 1-3mm thick; pocket of coarser sand in lower 10cm.
2m	2Ab2	87-104	10YR 4/2 3/1	SiL	m	vh	ai	V. common shells, few bone fragments; common vertical cracks 3-8mm wide filled w/ sand; zone of black, peaty material 0-5cm thick at base; C14 sample.
1	3C1b2	104-108	10YR 7/2 5/2	S	m	vh	ci	Indurated, discont. w/ contorted bedding & some sand from 3C2b2. Discont., laminated.
	3C2b2	108-115	10YR 7/2 6/2	S	sg	lo	aw	Faint bedding.
	3C3b2	115-140	10YR 8/1 6/2	S	sg	lo	aw	Contorted bedding locally common; indurated.
	3C4b2	140-149	10YR 7/1 5/2	S	m	vh	ai	
	3Btgb3	149+	10YR 8/2 6/2 5YR 5/6 5/6					Blackwater Draw Fm; heavily gleyed.

TABLE A1.1 CONT'D.

Bw-58 (core & blowout exposure)

Strat.	Horizon	Depth (cm)	Color Dry Moist	Text.	Str.	Cons.	Bndy.	Remarks
								Dune sand 1-2 m thick occurs locally; 2 layers (upper=10YR 6/2 moist, and lower =10YR 4/2 moist).
5s?	C	0-38	7.5YR 5/1 3/1	fSL	1msb	lo	ai	
3	Ab1	38-41	10YR 5/1 3/1	fSL	1csb	sh	ai	Locally up to 30cm thick.
	Cb1	41-67	10YR 7/2 5/2	LfS	sg	lo	ai	Sandy marl.
	A1b2	67-87	10YR 6/3 8/1 10YR 7/1 5/1	LfS	1msb	sh	cw	Sandy marl; common dark OM stains.
2s or 3s	A2b2	87-120	10YR 7/1 8/1 10YR 5/2 7/1	LfS	1msb	sh	ai	Common dark OM stains.
2d	2Cb2	120-137	10YR 5/1 3/1 10YR 7/1 5/1	---	m	vh	ai	Interbedded diatomite and diatomaceous mud; zone varies from 10-20cm thick where present; locally missing.
1	3Ab3	137-170	10YR 6/1 3/1	LfS	1msb	fr	ai	
	3C1b3	170-220	10YR 7/2 5/1	fS	sg	lo	ai	
B	4C2b3	220-225+	10YR 8/1 8/2	---	m	vh		Lacustrine carbonate.

TABLE A1.2. DESCRIPTION OF BAILEY DRAW CORE BW-39

Strat.	Horizon	Depth (cm)	Color Dry Moist	Text.	Str.	Cons.	Bndy.	Remarks
Fill		0-10						
4s	A	10-43	7.5YR 4/2 3/2	SCL	2msb	h	c	
	AB	43-62	7.5YR 3/2 3/2	CL	1fpr& 2mab	h	c	
	Bt1	62-92	7.5YR 4/2 3/2	CL	2fr& 3mab	h	c	Thin, discont. clay films.
	Bt2	92-102	10YR 3/2 7.5YR 3/2	CL	3fpr& 3mab	h	g	Thin, cont. clay films.
3c	A1b	102-122	7.5YR 3/2 2/2	SCL	2mab	h	g	C14 sample 102-143cm.
	A2b	122-143	7/5YR 3/2 2/1	SL	2mab	sh	g	C14 sample.
	ACb	143-163	10YR 5/2 4/2	SL	1msb	sh	c	
	C1b	163-184	10YR 5/2 7.5YR 3/2	SL	m	h	a	
1	2C2b	184-198		fgrSL				
	3C3b	198-208	10YR 6/1 5/1	SL	m	sh	a	
	4C4b	208-229		grSL				
	5Cgb	229-264	2.5YR 7/4 6/2	SCL	m	h	a	
	6Cb	264-290+	10YR 8/1 ---					Carbonate gravel & granules; weathered calcrete?

TABLE A1.3. DESCRIPTIONS OF BFI NORTH PIT AND SOUTH PIT SECTIONS

North Pit

Strat.	Horizon	Depth (cm)	Color Dry Moist	Text.	Str.	Cons.	Bndy.	Remarks
4s	A	0-16	--- 7.5YR 3/4	fSL	1msb	fr	gw	Some compaction; few f. bodies of carb.
	BA	16-40	--- 7.5YR 4/4	SCL	1mpr& 2msb	fr	gs	Few f. bodies & thr. of carb.
	Bk	40-86	--- 7.5YR 4/4	SCL	1msb	fr	gw	Com. f. bodies & thr. of carb.
	BAb	86-123	--- 10YR 3/4	SCL	1mpr& 1msb	fr	cs	Few 1-2mm pri. carb. bodies; cumulic A horizon.
3s	Ab	123-147	--- 10YR 3/3	SCL	1mpr& 1msb	fr	cs	Few 1-2mm pri. carb. bodies; C14 sample.
	Akb	147-184	--- 10YR 3/3	SCL	1mpr& 1msb	fr	cw	Few 1-2cm pri. carb. nodules; Stage I.
	Bkb	184-220+	--- 7.5YR 6/4	L	m	fr		Stage I. Buried dune?

TABLE A1.3 CONT'D.

South Pit

Strat.	Horizon	Depth (cm)	Color Dry Moist	Text.	Str.	Cons.	Bndy.	Remarks
Fill		0-55						
4s	Ak	55-75	--- 7.5YR 3/3	L	m	vh	aw	Stage If.
	Btk	75-125	--- 7.5YR 4/4	SCL	1cpr& 2msb	vh	cw	Stage If.
	BCk1	125-155	--- 7.5YR 4/6	SCL-	1csb	vh	cw	Stage I; finely divided carb.
	BCk2	155-180	--- 7.5YR 4/6	SL	1csb	vh	cw	Stage I; more finely divided carb. than BCk1.
3c	ABkb	180-205	--- 10YR 3/2	SL	1msb	vh	cw	Stage 1f; C14 sample.
	2Cb	205-245	--- 10YR 6/3	L	1fsb	h	aw	
B	3Ckb	245-280+	10YR 7/2 ---	SCL				Stage If; lacustrine clay?

TABLE A1.4. DESCRIPTION OF BLEDSOE CORE SU-1

Strat.	Horizon	Depth (cm)	Color Dry	Moist	Text.	Str.	Cons.	Bndy.	Remarks
5s?	C								Sand in dunes; up to 2m thick; exposures poor.
C	Btb1	0-50	7.5YR 5/4	4/4	SL	1mpr& 2msb	h	ai	Cont., thin clay films.
B	2Bkb1	50-75	10YR 8/3	7/3	SL	1msb	h	cs	Stage IIf; calcic horizon formed in marl.
	2Cb1	75-97	10YR 8/2	6/3	---	m	h	ai	Marl.
A	3Ab2	97-107	10YR 6/2	4/2	SL	1msb	sh	ai	Thickness varies from 5-15cm.
	3ACb2	107-124	10YR 6/3	5/3	LS	sg	lo	cs	
	3Cb2	124-164	7.5YR 7/2	6/2	fS	sg	lo	ci	
	4Btb3	164-181	5YR 5/6	5/4	SL	2mpr& 3msb	h	c	Cont., thick clay films; top of Blackwater Draw Fm. Stage If.
	4Btkb3	181-194	5YR 5/6	4/6	SCL	2mpr& 3msb	h	a	
	5C1b3	194-236	5YR 8/2	7/3	---	m	h	gr	Lake carbonate; few, faint mottles of 8/3 (d).
	5C2b3	236-326	5YR 8/3	7/3	---	m	h	a	Lake carbonate.
	6Btkb4	326-349	5YR 5/6	4/6	SCL	2mpr& 2msb	h	c	Stage IIf.
	7Cb4	349-360	10YR 8/2	8/2	CL	m	h	c	Lacustrine clay.
	8Btb5	360-370+	5YR 5/6	4/6	SCL	2mpr& 2msb	h		Cont., thick clay films.

TABLE A1.5. DESCRIPTION OF BOONE CORE MD-2

Strat.	Horizon	Depth (cm)	Color Dry	Moist	Text.	Str.	Cons.	Bndy.	Remarks
Fill		0-60							
5s	Bt	60-82	7.5YR 7/3	5/4	SL	1mpr& 2msb	h	a	Thin, discont. clay films.
	Bk	82-120	7.5YR 7/2	5/4	SL	m	fr	a	Stage IIf.
4s	Btkb1	120-200	7.5YR 7/4	5/4	L	1mpr& 2msb	h	c	Stage II.
3c	Cb1	200-215	7.5YR 8/2	10YR 6/2	SL	m	h	c	
	Ab2	215-265	10YR 6/2	4/2	SL	1msb	h	c	C14 sample.
	C1b2	265-400	10YR 8/1	7/2	SL	m	h	a	
1	2C2b2	400-425+	10YR 8/2	7/2	LmS	sg	lo		

TABLE A1.6. DESCRIPTION OF BROWNFIELD CORE SU-16

Strat.	Horizon	Depth (cm)	Color Dry	Moist	Text.	Str.	Cons.	Bndy.	Remarks
Fill		0-120							
4s	A	120-150	10YR 4/3	3/2	fSL	2msb	h	c	
	Bt	150-240	10YR 6/3	4/3	SL	2mpr& 2msb	h	c	
	Btk1	240-360	10YR 6/3	4/3	SL	2mpr& 2msb	h	c	Stage If.
	Btk2	360-430	10YR 5/3	4/2	SL	2mpr& 2msb	h	c	Stage If.
3c	Ab	430-455	10YR 5/2	4/2	SL	2msb	sh	c	C14 sample.
	C1b	455-480	10YR 7/2	5/2	LS	1msb	sh	c	
	C2b	480-540	10YR 8/2	6/2	LS	m	h	c	Discont. zone of gleyed, OM-rich mud 535-540cm.
	C3b	540-550	10YR 7/2	5/2	LS	m	h	g	
1	2C4b	550+	------------	---	fS	sg	lo		

TABLE A1.7. DESCRIPTION OF CANNON CORE BW-29

Strat.	Horizon	Depth (cm)	Color Dry	Moist	Text.	Str.	Cons.	Bndy.	Remarks
Fill		0-20							
4s	A	20-40	7.5YR 4/2	3/2	SiL	2msb	h	c	
	ABt	40-62	7.5YR 4/2	3/2	SiL	3msb	h	c	Thin, discont. clay films.
	Bt	62-104	7.5YR 4/2	3/2	SiCL	3msb	h	c	Thin, cont. clay films.
	Btk	104-118	7.5YR 5/2	4/2	CL	3msb	h	c	Stage IIf.
	Bt'	118-150	7.5YR 4/2	3/2	CL	2msb	h	c	
	Btk'	150-166	7.5YR 4/2	3/2	L	2msb	h	a	Stage If.
3c	Ab	166-189	10YR 3/3	3/2	CL	2mab	h	c	C14 sample.
	ACb	189-218	10YR 4/2	3/2	SL	1fsb	h	c	
	Cb	218-250	10YR 5/2	4/2	SL	1fsb	h	c	
1	2Cb	250-275	----------------	---	fS	sg	lo	a	
	3Cb	275-290+	----------------	---	grS	sg	lo		

TABLE A1.8. DESCRIPTIONS OF CLOVIS SECTIONS BW-60, 70, AND 72

Bw-60 (North Bank)

Strat.	Horizon	Depth (cm)	Color Dry	Moist	Text.	Str.	Cons.	Bndy.	Remarks
C	A	0-40	7.5YR 5/2	4/2	fSL	1csb	sh	cs	Description is composite of West and East walls.
	ABtk	40-90	7.5YR 6/4	5/4	fSL	2msb	h	cs	Stage If.
	Btk	90-130	7.5YR 7/2	10YR 5/3	fSL	1mpr& 2msb	h	ai	Stage If.
B	2C1	130-160	10YR 8/1	7/3	--	m	sh	ai	Marl.
A	3C2	160-195	10YR 7/3	6/3	S	sg	lo	ai	
	4Btkb	195+	5YR 6/4	5/4	fSL	2mpr& 2csb	vh		Stage IIf; thin, cont. clay films; top of Blackwater Draw Fm.

TABLE A1.8 CONT'D.

Bw-70 (Mayfield Dairy Road)

Strat.	Horizon	Depth (cm)	Color Dry	Moist	Text.	Str.	Cons.	Bndy.	Remarks
C	A	0-30	7.5YR 4/2	2/2	fSL	1mgr	sh	cs	
	Bt	30-85	7.5YR 6/3	4/3	fSL	1cpr& 2msb	sh	gs	Common, thin clay films.
B	Akb	85-128	7.5YR 6/2	4/2	fSL	1mgr	fr	cs	Stage If-;C14 sample.
	Bkb	128-140	10YR 7/2	6/2	fSL	1msb	fr	cs	Stage If; weathered marl?
	C1b	140-170	10YR 7/3	6/3	mLS	m	L	cs	Marl.
	C2b	170-200+	10YR 8/1	8/2	mLS	m	L		Marl.

TABLE A1.8 CONT'D.

Bw-72 (West Bank)

Strat.	Horizon	Depth (cm)	Color Dry	Moist	Text.	Str.	Cons.	Bndy.	Remarks
	Ap	0-15	10YR 4/2	---	LfS	sg	lo	cs	Sand sheet.
	C	15-30	10YR 5/3	---	fS	sg	lo	ai	Sand sheet.
C	Bt1b1	30-38	7.5YR 5/4	4/3	fSL	1mpr& 2msb	sh	cs	Thin, discont. clay films.
	Bt2b1	38-47	7.5YR 5/4	4/3	fSL	2mpr& 2csb	sh	cs	Thin, cont. clay films; some OM stains (ABt?).
	Bt3b1	47-57	7.5YR 5/3	3/3	fSL	2mpr& 2csb	sh	ai	Thin, cont. clay films.
B	2C1b1	57-65	10YR 8/1	7/2	---	m	sh	g	Lacustrine carbonate; locally missing.
	2C2b1	65-137	10YR 8/1	6/2	---	m	sh	as	Lacustrine carbonate; platy structure locally common; lower 1/2 less dense.
	3Akb2	137-150	10YR 6/1	4/1	LfS	1msb	sh	cs	Stage If; C14 sample from this horizon on North Bank (Bw-58).
A	3Cb2	150-157	10YR 7/2	6/2	LfS	sg	lo	cs	Alluvium?
	3Ab3	157-172	10YR 7/2	4/2	LfS	1msb	sh	cs	Top of Blackwater Draw Fm.
	4Btkg1b3	172-192	2.5YR 7/2 5YR 5/6	6/2 5/4	L	3cpr& 3csb	vh	cs	Stage If; 70% 2.5YR, 30% 5YR; thick, cont. clay films.
	4Btkg2b3	192-220	10YR 6/2 5YR 6/4	5/2 5/4	L	3cpr& 3csb	vh	cs	Stage If; 30% 10YR, 70% 5YR; 10YR more common near base; thick, cont. clay films.
	5Cb3	220-275	10YR 8/2	8/3	---	m	sh	ai	Lacustrine carbonate.
	6Btkgb4	275-350	5YR 6/6 10YR 6/2	5/6 5/2	fSL	3cpr& 3csb	sh	ai	Stage II; 50% 5YR, 50% 10YR; limonite stains common in upper 20cm; this zone penetrates another marl layer.

TABLE A1.9. DESCRIPTION OF DAVIS CORE BW-45

Strat.	Horizon	Depth (cm)	Color Dry	Moist	Text.	Str.	Cons.	Bndy.	Remarks
Fill		0-20							
4s	A	20-55	10YR 3/2	2/2	LS	1msb	sh	c	
	Bt	55-100	7.5YR 5/4	3/4	LS	1mpr& 2msb	sh	c	Thin, cont. clay films.
	C	100-146	7/5YR 6/4	4/4	S	1msb	sh	a	
3c	Ab1	146-185					sh	c	
	C1b1	185-250	7.5YR 6/2	4/2	SL	1msb	fr	a	
	C2b1	250-300	7.5YR 5/2	4/2	SL	m			
2m	Akb2	300-330	--------------	---	SL	2mab	vh	c	Stage If; C14 sample.
1	2C1b2	330-375	--------------	---	fS	sg	lo	a	
	3C2b2	375+	--------------	---	cS&gr	sg	lo		

TABLE A1.10. DESCRIPTION OF EDMONSON CORE RW-19

Strat.	Horizon	Depth (cm)	Color Dry	Moist	Text	Str.	Cons.	Bndy.	Remarks
Fill		0-75							
5s	A	75-120	10YR 5/3	3/3	SL	1msb	vh	c	
	Bt	120-145	10YR 5/3	3/3	SL	1mpr& 2msb	h	c	Thin, discont. clay films.
	Bw	145-175	10YR 6/3	4/3	LS	1msb	sh	a	
4s	Akb1	175-255	10YR 3/3	3/2	SL	2msb	h	c	Stage If.
	Btb1	255-300	10YR 6/3	4/3	SL	1mprd 2msb	sh	c	Thin, cont. clay films.
3c	C1b1	300-308	10YR 7/2	5/3	SL	1msb	h	c	
	2C2b1	308-311	--------------	---	fS	sg	lo	a	
	3Ab2	311-350	10YR 2/1	2/1	CL	2mpr& 2mab	vh	a	
	3C1b2	350-368	10YR 8/1	5/1	L	1msb	sh	c	
	3C2b2	368-374	10YR 7/2	5/3	SL	1msb	h	a	
2m	4C3b2	374-400	--------------	---	---	---	lo	a	Interbedded mud & olive sand; shell common 374-379cm; C14 sample 360-379cm.
1	4C4b2	400-420	--------------	---	---	---	lo	a	Interbedded tan sand and sandy mud; lenses=3-5cm thick.
	5C5b2	420+	--------------	---	---	---	lo		Interbedded sand & gravel.

TABLE A1.11. DESCRIPTION OF ENOCHS CORE YH-33

Strat.	Horizon	Depth (cm)	Color Dry	Moist	Text.	Str.	Cons.	Bndy.	Remarks
Fill		0-130							
4s	A	130-142	5YR 5/3	3/3	fSL	1msb	fr	c	
	Bt1	142-180	5YR 5/4	3/4	fSL	2msb	fr	c	Thin, discont. clay films.
	Bt2	180-215	5YR 5/3 7.5YR 3/2		fSL	1mpr& 1msb	fi	c	Thin, discont. clay films.
	BAt1	215-250	7.5YR 5/2	3/2	L	2msb	fi	c	Cumulic A horizon; thin, discont. clay films; C14 sample 210-274cm.
3c	BAt2	250-270	7.5YR 5/2 10YR 3/2		L	1msb	fi	a	Thin, discont. clay films.
	Ab	270-274	10YR 3/1 5YR 3/2		L	2mab	fi	c	
	C1b	274-300	10YR 7/1	5/2	CL	m	fi	c	
	2C2b	300-310	10YR 7/2	4/2	SL	m	fi	a	
1	3C3b	310-400+	------------	---	grS	sg	lo		

TABLE A1.12. DESCRIPTION OF EVANS CORE BW-17

Strat.	Horizon	Depth (cm)	Color Dry	Moist	Text.	Str.	Cons.	Bndy.	Remarks
Fill		0-55							
5s2	A	55-83	7.5YR 4/3	3/2	SCL	1msb	sh	cs	
	Bw	83-105	7.5YR 5/3	4/2	SCL	2msb	sh	cs	
	C	105-120	------------	---	S	m	sh	ai	Interbedded 5YR S & OM-rich SL.
5s1	Ab1	120-195	7.5YR 4/2	3/2	fSL	2msb	sh	cs	C14 sample.
	Cb1	195-205	7.5YR 5/3	4/2	S	sg	lo	ai	
3c	2A1b2	205-245	10YR 6/1	3/2	---	m	sh	cs	
	2A2b2	245-280	7.5YR 5/2	4/2	---	2msb	vh	cs	C14 sample.
	2A3b2	280-305	7.5YR 6/2	5/2	---	1msb	vh	cs	Marl mottled w/OM.
			7.5YR 5/2	4/2					
	2C1b2	305-360	7.5YR 6/2	5/2	---	m	sh	cs	
	2C2b2	360-410	7.5YR 8/2	4/2	---	m	sh	cs	
	2C3b2	410-455	10YR 8/1	6/1	---	m	sh		Auger 455-590cm; 3C to 590cm.
		550-590	10YR 6/1	5/1	---	m	sh		Marl sample off of auger.

TABLE A1.13. DESCRIPTION OF FLAGG CORE RW-11

Strat.	Horizon	Depth (cm)	Color Dry	Moist	Text.	Str.	Cons.	Bndy.	Remarks
5s	A	0-25	7.5YR 4/2	3/2	L	2mab	vh	c	
	Bt	25-72	10YR 4/2	3/2	SL	2mab	h	a	Thin, discont. clay films; faint bedding 42-65cm.
4s	Ab	72-104	10YR 3/6	2/2	L	2mab	h	c	C14 sample.
	ABt1b	104-145	10YR 4/3	3/2	SL	1mpr& 2msb	h	c	Thin, discont. clay films.
	ABt2b	145-164	10YR 4/3	3/2	SL	2mpr& 2msb	h	c	Thin, discont. clay films.
	Bt1b	164-195	7.5YR 5/4	4/2	SL	2mpr& 2msb	h	c	Thick, cont. clay films.
	Bt2b	195-255	7.5YR 5/4	3/4	SL	2mpr&2 msb	h	a	Thick, cont. clay films.
2d	C1b	255-270	--------------	---	SL	1mab	h	a	
	C2b	270-283	--------------	---	--	---	--	as	Interbedded mud, diatomite & carbonate; C14 sample.
2m	2C3b	283-384	--------------	---	--	---	--	as	Interbedded mud, olive clay & sand; oxidation mottles (10YR 5/6) 374-384cm; threads of silica 381-384cm; C14 sample 286-300cm.
1	3C4b	384-414	--------------	---	mS	sg	lo		Clean, well sorted, bedded mS.

TABLE A1.14. DESCRIPTIONS OF GIBSON CORES BW-31 AND BW-78

Bw-31

Strat.	Horizon	Depth (cm)	Color Dry	Moist	Text.	Str.	Cons.	Bndy.	Remarks
5s	C	0-50	10YR 7/2	4/2	S	sg	lo	c	Modern dune.
3c	2C1	50-100	10YR 7/1	4/1	S	m	lo	c	
	2C2	100-200	10YR 6/1	4/1	S	m	lo	g	
3s	3C3	200-270	10YR 6/2	4/2	S	sg	lo	a	
2d	4C4	270-370	10YR 6/1	4/1	--	m	fr	a	Interbedded diatomite (10YR) & mud (N). C14 sample 285-335cm.
1	5C5	370-470+	-------------	---	S	sg	lo		

TABLE A1.14 CONT'D.

Bw-78

Strat.	Horizon	Depth (cm)	Color Dry Moist	Text.	Str.	Cons.	Bndy.	Remarks
3s	C1	0-12	10YR 7/2 5/2	LS	sg	lo	ai	Clean sand interbedded w/ lenses of OM-rich sand; some OM-rich bodies 2-10mm dia. (bioturbation?)=10YR 5/1 & 4/1; snails locally common.
	C2	12-25	10YR 7/2 5/1	fSL	sg	lo	ci	Calcareous; snails common; slight OM enrichment in lower 2cm.
	Ck	25-70	10YR 7/2 5/2	fS	sg	lo	as	Stage If; subhorizontal laminae 2mm thick common in lower half of unit; lens of OM staining 1-3mm thick @ 50cm depth.
	Ab1	70-85	10YR 6/1 4/1	fSL	1mpr	fr	cs	Common OM stains (charcoal?) @ 70-75cm.
	C1b1	85-110	10YR 7/1 5/2	fSL	m	fr	cs	
	C2b1	110-171	10YR 7/1 5/1	fSL	m	fr	cs	
	Ck1b1	171-230	10YR 7/1 5/2	fS	m	fr	cs	Stage If.
	Ck2b1	230-265	10YR 8/1 6/2	fS	m	fr	ai	Stage If.
	Ck3b1	265-273	10YR 7/1 5/1	fS	m	fr	ai	Stage If; common OM stains & few charcoal flecks.
	A1b2	273-280	10YR 6/1 4/1	fS	m	fr	cs	Common flecks & stains of charcoal; C14 sample.
	A2b2	280-354	10YR 5/1 4/2	SL	m	fr		Flint flakes in situ @285 & 295cm; burned bone fragments throughout; flakes from screeing throughout; few silicified plant remains throughout; burned caliche @ 305cm.
2d								Auger: clean, white sand 354-400cm, diatomite 400-407cm. Noncalc. below 280cm.

TABLE A1.15. DESCRIPTION OF GLENDENNING SECTION MU-3

Strat.	Horizon	Depth (cm)	Color Dry	Moist	Text.	Str.	Cons.	Bndy.	Remarks
Fill		0-14						cw	Historic slopewash.
4s	Ap	14-42	7.5YR 5/3	4/3	fSL	m-1msb	vh	gs	Compacted due to plowing.
	BA	42-81	7.5YR 6/3	4/4	fSL-	1msb	sh	gs	Few 0.5-2cm pri. carb. nodules.
	Bw1	81-108	7.5YR 7/3	4/4	fSL-	1msb	sh	gs	Few-com. 0.5-3cm pri. carb. nodules.
	Bw2	108-137	7.5YR 7/3	4/4	LfS	1msb	sh	gs	Com. 0.5-5cm pri. carb. nodules.
	2Bw3	137-170	7.5YR 7/3	5/4	LfS	1msb	sh	ai	Com.-many 0.5-5cm pri. carb. nodules; few up to 20cm; many burned; peds from Ab (below) mixed in.
3c	2Akb	170-213	10YR 4/2	3/2	LmS	m-1fsb	h	ci	Com. faint carb. films & thr. on ped faces; C14 sample.
	2ACgkb	213-230	10YR 6/3 10YR 7/2	5/2 6/3	L	1fsb	h	cw	Com.-many carb. films & thr. on ped faces.
	2Cgkb	230-263	2.5YR 8/2	7/3	SiL	m-1fsb	h	ci	Few-com. thr. of carb.
	3Cgb	263-298	2.5YR 8/2	7/3	SiL	m	h	ci	Com. 1-2cm pri. carb. nodules; some larger ones in pockets.
	4Cgb	298-328	2.5YR 8/2	7/3	SiL	m	h	ci	Few pockets of 10-20cm carb. nodules.
1	5Cb	328-360	2.5YR 8/2	7/3	gr m-cS	sg	lo	ai	Com. 1-10cm rounded pri. carb. nodules; mussell shells v.com.

Snails common below 170cm; bone frags. com. below 230cm. |

TABLE A1.16. DESCRIPTION OF HALSELL CORE BW-27

Strat.	Horizon	Depth (cm)	Color Dry	Moist	Text.	Str.	Cons.	Bndy.	Remarks
Fill		0-20							
4s	ABt	20-35	7.5YR 4/2	3/2	SCL	2mab	vh	c	
	Bt	35-80	7.5YR 5/4	4/2	SL	2msb	vh	c	
	BC	80-100	7.5YR 6/2	4/2	SL	1msb	vh	a	
3c	Ab	100-115	10YR 4/2	3/2	CL	2mab	vh	c	C14 sample.
	C1b	115-150	10YR 7/1	5/2	SCL	m	vh	g	
	C2b	150-212	10YR 7/2	6/2	SL	m	eh	a	
1	2C3b	212-218+	----------------	---	fS	sg	lo		

TABLE A1.17. DESCRIPTIONS OF HOUCK CORES RW-2, RW-3

Rw-2

Strat.	Horizon	Depth (cm)	Color Dry	Moist	Text.	Str.	Cons.	Bndy.	Remarks
Fill		0-50							
4s	A	50-85	7.5YR 5/4	4/2	SL	2msb	sh	c	Thin, cont. clay films.
	Bt	85-128	7.5YR 5/4	3/4	SL	1mpr& 2msb	sh	c	
	C1	128-146	7.5YR 5/4	4/4	LS	m	sh	a	
	2C2	146-155			grSL	sg	lo	a	Non-gravelly matrix=C1.
	3C3	155-210	7.5YR 6/4	4/2	SL	m	sh	a	
	3C4	210-230	7.5YR 5/4	3/4	SL	m	sh	a	
	3C5	230-250	7.5YR 6/4	4/4	SL	m	sh	a	
2?	3Ab	250-285	7.5YR 3/2	3/2	SCL	2mab	vh	c	
1	3C1b	285-320	7.5YR 4/4	3/2	SL			a	
	3C2b	320-335	--------------	---	mS	sg	lo	a	
	3C3b	335-420	--------------	---		sg	lo	a	Fining upward sequence: 420-410cm=mS, 410-375cm=fS, 375-335cm=LfS.
	3C4b	420-450	--------------	---	fSCL	sg	lo	a	
	4C5b	450-480+	--------------	---	grLS	sg	lo		

TABLE A1.17 CONT'D.

Rw-3

Strat.	Horizon	Depth (cm)	Color Dry	Moist	Text.	Str.	Cons.	Bndy.	Remarks
Fill		0-60							
4s	A	60-110	7.5YR 4/2	3/2	SL	1msb	sh	c	
	Btk1	110-220	7.5YR 5/4	4/2	L	1mpr& 2msb	h	c	Stage If; thin, cont. clay films.
	Btk2	220-270	7.5YR 5/2 & 5/4	4/2	L	1mpr& 2msb	h	c	Stage If+; thin, cont. clay films.
	Btk3	270-320	7.5YR 4/4	3/2	SiL	1mpr& 2msb	sh	c	Stage IIf; thin, cont. clay films.
	Btk4	320-360	7.5YR 5/2 & 5/4	3/2	SiL	1mpr& 2msb	sh	g	Stage If; thin, cont. clay films.
3c	Ab	360-418	7.5YR 5/2	3/2	L	2msb	sh	c	Has characteristics of Btk4; C14 sample.
	2C1b	418-490	7.5YR 7/2	5/4	LS	1fsb	sh	c	
	2C2b	490-530+	10YR 8/1	5/2	LS	1fsb	sh		Few oxidation mottles 450-455cm.

TABLE A1.18. DESCRIPTION OF JORDE SECTION BW-93

Strat.	Horizon	Depth (cm)	Color Dry	Moist	Text.	Str.	Cons.	Bndy.	Remarks
B	Bk	0-20	10YR 7/2	6/1	---	1msb	sh	cs	Marl w/common carb. concr.
	C	20-32	10YR 7/2	6/1	---	m	sh	cs	Marl.
	A1b1	32-62	10YR 4/2	3/1	SiL	2msb	vh	cs	V. few, faint carb. films; C14 sample, 32-42 cm.
A	2A2b1	62-87	2.5Y 4/2	3/1	SL	2msb	vh		Alluvium?
	3Btgkb2	87+	10YR 8/2 5YR 5/6	6/2 5/6			vh		Blackwater Draw Fm; heavily gleyed.

TABLE A1.19. LUBBOCK LAKE SITE, COMPOSITE STRATIGRAPHIC SECTION*

Stratum	Valley-axis facies	Valley-margin facies
5	Stratum 5m2: up to 1m thick; gray to very dark gray (5YR 5/1 to 3/1, dry); clay; weakly stratified.	Stratum 5s2, 5g2: 10-25cm thick; brown (for example, 7.5YR 5/3, dry); sandy clay loam to sandy loam interbedded with common sand and gravel lenses.
	Stratum 5m1: same as 5m2.	Stratum 5s1, 5g1: 30-75cm thick; brown (for example, 7.5YR 5/3, dry); sandy clay loam to sandy loam interbedded with few sand and gravel lenses.
4	Stratum 4m: same as 5m2.	Stratum 4: 1-3+m thick; brown (for example, 7.5YR 5/4, dry); sandy clay loam to sandy loam.
	Stratum 4s & 4m: <1m thick; olive gray (2.5YR hues); laminated to massive, often cross-bedded, well sorted, loamy fine sand to sandy clay loam interbedded with blocky to granular, somewhat more organic clay to clay loam.	No valley margin equivalent of 4s & 4m.
3	Stratum 3c: 30-100+cm thick; highly calcareous; white (10YR 7/1, dry); massive to platy, friable, silty clay to silty clay loam.	Stratum 3s: 30-100+cm thick; light brown (7.5YR 7/3, dry); sandy loam.
2	Stratum 2s: up to 30cm thick; light gray (for example, 2.5Y 7/2, dry); sandy loam.	
	Stratum 2m: 30-80cm thick; gray (for example, 10YR 5/1, dry); loam to silty clay loam to clay; locally abundant silicified roots; few lenses of diatomite.	Stratum 2s (facies of 2A and 2B): up to 2m thick; gray (for example, 2.5Y 7/2, dry); silt clay interbedded with light gray (for example, 2.5Y 7/2, dry) sandy clay; or pale brown (for example, 10YR 6/3, dry); sandy clay loam.
	Stratum 2d: 30-100cm thick; light gray (10YR 7/1, dry) diatomite interbedded with gray (for example, 10YR 5/1, dry) silt to clay.	

*Modified from Holliday, 1985c, Table 2.
Note: facies are not necessarily time equivalent. No generalized description of stratum 1 was possible due to high variability.

TABLE A1.20. DESCRIPTIONS OF LUBBOCK LANDFILL SECTIONS C-3, D-4, E-4, AND W-4

C-3

Strat.	Horizon	Depth (cm)	Color Dry	Moist	Text.	Str.	Cons.	Bndy.	Remarks
5s	A/B	0-25	7.5YR 3/3	---	SCL	1msb	sh	cs	Historic slopewash.
3s	Ab	24-40	7.5YR 3/2	---	SCL	1msb	h	cs	
	ABkb	40-60	7.5YR 4/3	---	SCL	1msb	h	cs	Stage If.
	C1b	60-92	7.5YR 6/4	---	SL	m	lo	cs	
	C2b	92-96	10YR 7/2	5/2	SL	m	sh	as	
1	2C3b	96-135	10YR 6/4	---	fS	sg	lo	aw	
	3C4b	135-180	------------	---	cgr			aw	
	4C5b	180-210+	5YR 6/6	4/6	SC	1msb	fr		Ogallala Fm.

TABLE A1.20 CONT'D.

D-4

Strat.	Horizon	Depth (cm)	Color Dry	Moist	Text.	Str.	Cons.	Bndy.	Remarks
	C	0-23	7.5YR 3/4	---	SL	1msb	h	cs	Historic slopewash.
5s	Ab1	23-45	7.5YR 3/2	---	SL	1msb	h	gs	
	Bwb1	45-74	7.5YR 3/3	---	SL	1msb	h	cs	
4s	Akb2	74-100	7.5YR 3/2	---	L	2msb	h	gs	Stage If; C14 sample.
	Btk1b2	100-125	7.5YR 4/3	---	L	1mpr& 2msb	h	gs	Stage If; thin, cont. clay films.
	Btk2b2	125-150	7.5YR 4/4	---	SiL	1mpr& 2msb	h	cs	Stage If+; thin, cont. clay films.
3si	Akb3	150-225	7.5YR 3/3	---	SiL	1mpr& 2msb	h	cs	Stage If; C14 sample.
	Btkb3	225-280	7.5YR 4/3	---	SL	1mpr& 2msb	h	as	Stage If; thin, discont. clay films.
3c	Ab4	280-310	10YR 3/2	---	SL	m	vh	as	C14 sample.
	C1b4	310-380	10YR 3/3	---	LS	m	vh		
1	2C2b4	380-400+	------------	---	S	sg	lo		

TABLE A1.20 CONT'D.

<u>E-4</u>

Strat.	Horizon	Depth (cm)	Color Dry	Moist	Text.	Str.	Cons.	Bndy.	Remarks
		0-210							Valley Fill.
									210-580cm = Blackwater Draw Fm.
	ABtkb1	210-228	5YR 3/3	3/2	SCL	1mpr& 3msb	vh	cs	Stage I-; common thin clay films.
	Btk1b1	228-255	5YR 3/4	3/3	SCL	1mpr& 3msb	vh	cs	Stage I-; cont. thick clay films.
	Btk2b1	255-305	5YR 3/4	3/3	SCL	3cpr& 3csb	vh	cs	Stage II+; cont. thick clay films.
	Btk3b1	305-350	5YR 4/4	3/4	SCL	3cpr& 3csb	vh	cs	Stage II+; cont. thick clay films.
	Btk4b1	350-410	5YR 5/4	4/4	SCL	3cpr& 3csb	vh	cs	Stage II+; cont. thick clay films; 1cm thick lens of bedrock rubble @ 410cm.
	2Bkb1	410-445	7.5YR 5/4	3/4	fS	sg	h	aw	Stage II+; sandy facies of rubble lens.
	3C1b1	445-450	------------	---	---	sg	h	as	Coarse rubble from bedrock.
	4C2b1	450-474	------------	---	fS	sg	lo	aw	
	5C3b1	474-494	------------	---	---	sg	lo	aw	Interbedded sand and coarse rubble from bedrock.
	6C4b1	494-535	------------	---	fS	sg	lo	aw	Few lenses of cS and clay 1-2cm thick; some planar crossbedding.
	7C5b1	535-540	------------	---	---	sg	lo	aw	Discont. lens of poorly sorted coarse rubble from bedrock.
	8C5b1	540-580	------------	---	fS	sg	lo	aw	Planar crossbedded w/few, discont. clay lenses up to 1cm thick.
	R	580+	------------	---	---	---	--	--	Blanco Fm.

TABLE A1.20 CONT'D.

W-4

Strat.	Horizon	Depth (cm)	Color Dry	Moist	Text.	Str.	Cons.	Bndy.	Remarks
5s	A	0-30	10YR 4/3	7.5YR 3/2	LS	1mgr	sh	cs	Lens of carb. gravel 1cm thick at base.
4s2	Akb1	30-65	10YR 5/3	3/2	SL	1msb	h	cs	Stage If; compacted.
	Btk1b1	65-105	7.5YR 5/4	4/4	SL	1mpr& 2msb	h	cs	Stage If; common, thin clay films.
	Btk2b1	105-155	7.5YR 5/4	4/4	SL	2mpr& 2msb	uh	cs	Stage IIf; continuous, thin clay films; pri. carb. nodules 2-5mm dia.
	Bwb1	155-174	7.5YR 5/4	4/4	LS	1msb	h	as	Pri. carb. nodules 1-3mm dia. common.
4s1	ABkb2	174-245	7.5YR 6/4	5/4	SL	1mpr& 2msb	h	cs	Stage IIf.
	ACb2	245-320	10YR 6/4	4/3	SL	1msb	so	cs	Pri. carb. nodules 1-5mm dia. common.
3s2	A1b3	320-393	10YR 6/3	4/2	SL	1msb	sh	aw	Discontinous lens of carb. gravel @ 381-393cm; cumulic A horizon.
3c	Ab4	393-450	10YR 6/2	4/1	L	2mgr& 1mpr	so	cs	
	C1b4	450-473	10YR 7/1	3/2	L	m	so	as	Discont. lens of sand 460-473cm.
2m	2C2b4	473-495	10YR 6/2	3/1	L	m	h	as	Discont. lenses of silica 1-5mm thick @ 483cm.
2d	3C3b4	495-520	-------------	---	---	m	h	as	OM-rich mud interbedded with sand (495-500cm); diatomite (500-510cm); mostly sand (510-520cm).
1	4C4b4	520-615+	-------------	---	---	----	---	---	Fine sand 520-565cm; fs w/ contorted bedding 565-582cm; fs w/gravel 582-615cm; lse. gravel 615+cm.

TABLE A1.21. DESCRIPTION OF LUPTON CORE YH-1

Strat.	Horizon	Depth (cm)	Color Dry	Moist	Text.	Str.	Cons.	Bndy.	Remarks
B	A	0-30	10YR 4/3	4/4	SL	1msb	h	c	
	AC	30-50	10YR 6/2	5/2	SL	1msb	h	c	
	C1	50-105	10YR 8/1	7/1	SL	m	h	c	
	C2	105-130	10YR 4/2	3/2	SL	m	h	c	
	Ck	130-140	10YR 8/1	7/1	SL	m	h	a	Common carb. nodules.
	Ab	140-160	10YR 6/2	5/2	SL	m	h	c	C14 sample.
	C1b	160-170	10YR 7/1	6/2	SL	m	fr	c	
	2C2b	170-275+	10YR 8/1	8/1	C	m	h		Blanco Fm? Entire section=lake sediments.

TABLE A1.22. DESCRIPTION OF MIDLAND SECTIONS MN-19, AND MN-5

Mn-19 (Locality 3W)

Strat.	Horizon	Depth (cm)	Color Dry	Moist	Text.	Str.	Cons.	Bndy.	Remarks
5s	C	0-15	7.5YR 7/4	5/3	fS	sg	lo	c	
4s	Ab1	15-55	7.5YR 4/6	3/6	fS	1msb	fr	c	
	Bw1b1	55-100	5YR 5/8	4/6	fS	1msb	fr	c	
	Bw2b1	100-170	5YR 6/8	4/8	fS	2msb	fr	c	
	Btb1	170-180	5YR 6/8	4/8	fS	2msb	fr	c	Few 1-3mm clay bands = 5YR 5/8.
?	Ab2	180-250	5YR 6/6	4/6	fS	1msb	fr	c	
	Btb2	250-300	5YR 5/8	4/6	fS	2msb& 1mpr	fr	c	Thin, patchy clay films.
	Bwb2	300-350+	5YR 6/8	4/8	fS	2msb	fr		Cross-bedding weakly preserved.

TABLE A1.22 CONT'D.

Mn-5

Strat.	Horizon	Depth (cm)	Color Dry	Moist	Text.	Str.	Cons.	Bndy.	Remarks
4s	A	0-5	7.5YR 5/4	4/4	fS	sg	lo	cs	
	ABw	5-15	7.5YR 5/6	4/6	fS	1msb	so	cw	Mixed A & Bw.
	Bw1	15-50	7.5YR 6/6	5/6	fS	1msb	so	g	
	Bw2	50-96	7.5YR 6/6	5/4	fs	1msb	so	cw	
mix	C1	96-122	10YR 7/3	6/3	fS	sg	so	cw	Mottled; common zones of 7.5YR 5-10cm diam.; 7.5YR more common 96-105cm.
			7.5YR 6/4	5/4					
3s1	C2	122-145	7.5YR 7/4	6/4	fS	sg	lo	cw	Heavily mottled; mottled zones commonly 5-10cm diam.; common pockets of white (7.5YR 8/2d) calc. sand.
			7.5YR 7/4	6/6					
			10YR 5/6	4/3					
			10YR 6/8	6/6					
1	C3	145-150+	7.5YR 8/2	7/4	LS	m	h		Note: 4s = "Red Sand"? 3s = "Gray Sand"? 1 = "White Sand"

TABLE A1.23. DESCRIPTION OF MUSTANG SPRINGS TRENCH 5

Strat.	Horizon	Depth (cm)	Color Dry	Moist	Text.	Str.	Cons.	Bndy.	Remarks
5s	Ap	0-26	7.5YR 4/3	3/3	L			aw	
	A	26-62	7.5YR 4/3	3/3	L	2msb	fr	aw	
4s	2Bt1	62-110	7.5YR 6/3	4/3	L	2msb	fr	d	
	2Bt2	110-126	7.5YR 5/4	4/3	SL	2msb	fr	aw	
3c	3C1	126-175	7.5YR 7/2	6/2	CL	1fsb	vfr	aw	Com. krotovinas, ave. 10cm diam.
	3C2	175-185	------------	4/2	CL	m	vh	as	
2d	4C3	185-243	------------	---	---	---	---	as	Interbedded diatomite and mud.
1	5C4	243+	------------	---	vfS				

TABLE A1.24. DESCRIPTIONS OF PAYNE CORES YH-6 AND YH-7

Yh-6

Strat.	Horizon	Depth (cm)	Color Dry	Moist	Text.	Str.	Cons.	Bndy.	Remarks
Fill		0-30							
5m	ABt	0-76	5YR 4/3	3/3	L	1msb	fr	a	Thin, discont. clay films.
3c	Ab	76-125	10YR 3/3	2/1	SiC	2msb	fr	c	
	C1b	125-330	7.5YR 6/2	5/2	grSL	m	fr	c	
1	2C2b	330+	--------------	---	grS	sg	lo		

TABLE A1.24 CONT'D.

Yh-7

Strat.	Horizon	Depth (cm)	Color Dry	Moist	Text.	Str.	Cons.	Bndy.	Remarks
Fill		0-70							
5s	ABw	70-115	7.5YR 4/3	3/3	SCL	1msb	h	c	Cumulic A horizon.
	C	115-145	5YR 4/1	3/1	SL	2msb	fr	c	
4s	Bwb	145-175							
	Btb	175-210	7.5YR 6/2	4/2	SL	1msb	h	c	
	Btkb	210-290	7.5YR 6/4	4/4	SL	2msb	h	c	Stage IIf.
3c	2C1b	290-370	10YR 7/2	5/3	L	m	h	c	
	2C2b	370-470	10YR 7/1	5/2	L	m	h	c	
1	3C3b	470-475+	----------------	---	Sgr	sg	lo		

TABLE A1.25. DESCRIPTION OF PLAINS PAVING NORTH WALL SECTION

Strat.	Horizon	Depth (cm)	Color Dry	Moist	Text.	Str.	Cons.	Bndy.	Remarks
4s	Ap	0-16	10YR 3/2	2/2	CL	1msb	fr	as	
	ABt	16-59	10YR 3/1	2/1	CL	1mpr& 2msb	fr	cs	Thin, discont. clay films.
	2BAt	59-77	7.5YR 4/2	2/2	fSL	1mpr& 2msb	fi	cs	Thin, discont. clay films.
	2Bt	77-96	7.5YR 5/4	4/4	fSL	2msb	fi	cs	Thin, cont. clay films.
	2Bw	96-123	7.5YR 6/3	4/4	fSL	1msb	fi	gs	
3c	2CB	123-134	10YR 8/2	5/4	fSL	1msb	fi	cw	
	2C1	134-162	10YR 7/3	6/4	LfS	m	fr	cw	
2m	3C2	162-172	10YR 4/3	3/3	LS	sg	lo	cw	discont., C14 sample.
1	3C3	172-185	10YR 7/3	5/4	fS	sg	lo	cw	Sandy mud occurs locally between 2C1 & 3C2.
	4C4	185+	----------------	---	grfS	---	--		

TABLE A1.26. DESCRIPTION OF PLAINVIEW SECTIONS S-2, S-3, S-7, AND S-12

S-2

Strat.	Horizon	Depth (cm)	Color Dry	Moist	Text.	Str.	Cons.	Bndy.	Remarks
5m	A	0-18	------------ 10YR 2/2		SiL	2fab	fr	cs	
	AB	18-34	------------ 10YR 2/1		SiL	2mab	fr	cs	C14 sample.
4si	BAt	34-46	7.5YR 3/2	2/2	SiL	3mab	h	cg	Thin, cont. clay films.
	Bt	46-87	7.5YR 3/2	2/2	SiL	2mpr& 3mab	vh	cw	Thin, cont. clay films.
4s	Btk1	87-100	7.5YR 4/2	3/2	SCL	1mpr& 2mab	vh	cw	Stage If; thin, discont. clay films.
	2Btk2	100-117	7.5YR 4/4	3/4	SL	1cpr	h	cs	Stage If-; thin, discont. clay films.
	2Bt	117-157	7.5YR 4/2	3/2	SL	1cpr	fr	cs	Stage If-; V. thin discont. clay films.
1	3Bw	157-185	10YR 5/4	4/3	LfS	1mpr	fr	cw	
	3C1	185-213	2.5Y 7/2	6/2	LvfS	m	vfr	aw	
	4C2	213-227	2.5Y 7/2	6/2	fS	m	lo	aw	Weakly bedded.
	4C3	227-273	10YR 8/2	6/2	fS	sg	lo	aw	Bedded.
	5C4	273+	10YR 8/2	6/2	grfS	sg	lo		Bedded.

TABLE A1.26 CONT'D.

S-3

Strat.	Horizon	Depth (cm)	Color Dry	Moist	Text.	Str.	Cons.	Bndy.	Remarks
5sl	A	0-24	10YR 3/2	2/2	SiL	1mpr& 2fab	fr	cs	
	BA	24-48	10YR 3/2	2/2	SiCL	2mpr& 2mab	vh	cs	
	Bt	48-70	10YR 3/3	2/3	SiC	2mpr& 3mab	vh	cw	Thin, cont. clay films.
	BAb	70-95	10YR 3/2	2/2	SiC	2mab	fr	cw	
4s	Ab	95-135	------------ 10YR 2/2		SiL	2msb	vfr	cw	C14 sample.
	Btkb	135-175	------------ 7.5YR 3/2		CL	1mpr& 2msb	fr	gs	Stage If; thin, cont. clay films.
	Bkb	175-210	10YR 6/3	4/3	L	1msb	vh	aw	Stage If.
3c	2Cb	210-248	10YR 8/1	2/5Y 6/2	SL	3fab	h	aw	
1	3C2b	248-266	------------ 2.5Y 6/2		LS	m	fr	aw	
	4C3b	266-288	10YR 7/3	6/2	LS	sg	l	aw	
	5C4b	288-303	------------ 2.5Y 6/2		SiL	m	S	aw	
	6C5b	303-325	10YR 7/3	6/2	S	sg	l	aw	5% gravel; shell common.
	7C6b	325-380	2.5Y 6/2	5/2	L	m	vh	aw	Shell common.
	8C7b	380+			vgrs	sg	l		

TABLE A1.26 CONT'D.

S-7

Strat.	Horizon	Depth (cm)	Color Dry	Moist	Text.	Str.	Cons.	Bndy.	Remarks
4s	A	0-45	10YR 5/3	3/2	L	1msb	h	cs	
	Bt	45-95	10YR 5/3	3/3	L	1mpr& 2msb	h	cs	Thin, discont. clay films.
3s	Akb1	95-135	10YR 4/2	2/2	L	1msb	h	gs	Stage If; C14 sample.
	Bk1b1	135-200	10YR 6/3	5/3	L	1mpr& 2msb	h	gs	Stage If.
	Bk2b1	200-245	10YR 6/3	5/3	L	2mpr& 2msb	h	cs	Stage IIf.
1m	Akb2	245-280	10YR 5/3	4/3	SL	2mpr& 2msb	h	gs	Stage If; C14 sample.
	Ab2	280-345	10YR 5/3	4/3	SL	2mpr& 2msb	h	aw	Snails common; 1-3cm dia. pri. carb. nodules & clasts of Blackwater Draw Fm common at base; C14 sample.
	K	345+							Ogallala Fm. calcrete.

TABLE A1.26 CONT'D.

S-12

Strat.	Horizon	Depth (cm)	Color Dry	Moist	Text.	Str.	Cons.	Bndy.	Remarks
Spoil		0-75							From quarry excavation.
5s2	A	75-100	10YR 3/2	2/2	CL	1mpr& 1fab	fr	cs	
	Bt	100-125	10YR 3/3	2/3	CL	2mpr& 3mab	vh	cs	Thin, cont. clay films.
5g	2Bw	125-145	10YR 3/4	4/4	gr SL	1mab	h	cw	Com. 1-3cm pri. carb. nodules.
4s	3Ab1	145-170	---------- 10YR 2/2		CL	2msb	vh	cw	C14 samples.
	3Btk1b1	170-200	---------- 10YR 3/2		CL	1mpr& 2msb	fr	gs	Stage If; cont. clay films.
	3Btk2b1	200-238	10YR 6/3	4/3	CL	1msb	vh	aw	Stage 1f+; thin, cont. clay films.
3c	3Cb2	238-255	10YR 8/1	7/2	SiCL	3fab	h	aw	Highly calc.; thin A horizon preserved locally.
2m	3Cb3	255-275	10YR 4/3	3/3	L	m	fr	aw	Contorted, disc. beds of OM-rich seds., varying 1-25cm thick. C14 sample.
1	3Cb4 4Cb5	275-414 414+	10YR 7/3	6/2	mS	sg	lo	aw	Gastropods common. Carb. gravel, 1-5cm diam.

TABLE A1.27. DESCRIPTION OF PROGRESS DRAW CORE BW-42

Strat.	Horizon	Depth (cm)	Color Dry	Moist	Text.	Str.	Cons.	Bndy.	Remarks
Fill		0-64							
4s	A	64-90	7.5YR 3/2	2/2	SL	2mab	h	c	
	ABt	90-117	7.5YR 4/2	3/2	SL	1mpr& 2msb	vh	g	Thin, discont. clay films.
	Btk1	117-174	7.5YR 4/2	3/2	SL	3fpr& 3fsb	vh	c	Stage If.
	Btk2	174-190	7.5YR 5/2	3/2	SL	3fpr& 3fsb	vh	a	Stage IIf.
2m	Ak1b	190-240	7.5YR 4/2	3/2	L	3fpr& 3fsb	vh	g	Stage If. C14 sample 190-275cm.
	Ak2b	240-285	10YR 3/2	7.5YR 3/2	L	3fpr& 3fsb	vh	c	Stage If.
	Ak3b	285-330	10YR 5/1	4/2	L	3fpr& 3fsb	vh	a	Stage IIf. C14 sample 275-330cm.
1	2Cg1b	330-345	2.5YR 7/2	4/2	grC	m	h	a	
	3Cg2b	345-355	10YR 4/2	3/2	C	m	h	a	30% mottles of 2.5Y; carb. coatings locally common.
	4Cb	355+	------------	---	grfS	sg	lo		

TABLE A1.28. DESCRIPTION OF QUINCY STREET SECTION RW-26

Strat.	Horizon	Depth (cm)	Color Dry	Moist	Text.	Str.	Cons.	Bndy.	Remarks
5s	A	0-45	7.5YR 4/2	2/2	fSL	1msb	sh	g	
	Bw	45-105	7.5YR 6/2	4/2	fSL	1fsb	sh	ai	
	2C1	105-150	7.5YR 7/2	4/2	fSL	m	h	ai	Common, discont. lenses of clay 1-5cm thick.
	3C2	150-175	7.5YR 7/2	5/2	SCL	m	h	ai	Laminated; common, discont. lenses of LfS.
	4C3	175-250	------------	---	---	m	h	ai	Complex interbedding of fS, mS & SC; locally convoluted w/some diapirs; some beds OM-rich.
	5C4	250-275	7.5YR 4/2	2/2	L	m	vh	ai	Lens of carb. gravel (1-5cm dia.) @ 260-265cm; C14 sample @ 265-275cm.
1	6C5	275-300+	------------	---	gr	sg	lo		Carb. gravel (1-5cm dia.).

TABLE A1.29. DESCRIPTION OF SEMINOLE-ROSE SECTION SE-13

Strat.	Horizon	Depth (cm)	Color Dry	Moist	Text.	Str.	Cons.	Bndy.	Remarks
	C	----	7.5YR 5/4	3/4	S	sg	lo	ai	Dune sand up to 1m thick.
C	Ab1	0-35	10YR 5/3	4/3	LS	1msb	sh	cs	
	Bwb1	35-100	10YR 7/3	6/4	LS	3cpr& 1csb	sh	cs	
	Bkb1	100-180	7.5YR 8/2 10YR 6/3	--- ---	LS	2cpr& 1csb	sh	cs	Stage II.
B	Ab2	180-215	10YR 7/2	5/3	LS	m	sh	cs	C14 sample.
	Ckb2	215-240	10YR 8/1	7/2	SL	m	h	cs	Common concr. & horizontal laminae of carb. in lower 5cm.
	2C1b2	240-295	10YR 7/3	6/3	S	sg	lo	cs	
	3C2b2	295-330	10YR 8/1	7/2	---	m	vh	cs	Massive carbonate.
	4C3b2	330-360	10YR 8/1	7/3	---	pl	vh	cs	Platy carbonate.
	5C4b2	360-390+	10YR 8/1	8/1	---	m	vh	ai	Massive carbonate.
	6Bkb3	390+	5YR 7/3	6/4	SL	2csb	eh	ai	Stage If+; Blackwater Draw Fm. ? up to 1.5m thick in broad pockets in 7Kb4.
	7Kb4	390+	10YR 8/1 7.5YR 8/2		---	m	eh		Ogallala Fm. calcrete. Entire section=lake sediments.

TABLE A1.30. DESCRIPTION OF SUNDOWN CORE SU-5

Strat.	Horizon	Depth (cm)	Color Dry	Moist	Text.	Str.	Cons.	Bndy.	Remarks
Fill		0-41							
5m	ABt	41-115	10YR 3/1	2/1	L	1mpr& 2msb	eh	c	Thin, discont. clay films.
4s	BAt	115-158	7.5YR 3/2	2/2	L	2mpr& 2msb	eh	c	Thin, cont. clay films; C14 sample.
	Btk1	158-213	7.5YR 5/4	3/4	L	2mpr& 3msb	vh	c	Stage IIf; thick, cont. clay films.
	Btk2	213-284	7.5YR 5/4	3/4	L	2mpr& 2msb	vh	a	Stage If; thin, discont. clay films.
3c	Ab	284-310	7.5YR 6/2	5/2	L	2mab	sh	c	C14 sample.
	C1b	310-375	10YR 8/1	5/2	L	m	sh	a	
1	2C2b	375-395+	------------	---	Sgr	sg	lo		

TABLE A1.31. DESCRIPTION OF TOLK CORE BW-36

Strat.	Horizon	Depth (cm)	Color Dry	Moist	Text.	Str.	Cons.	Bndy.	Remarks
Fill		0-20							
5m	A	20-50	10YR 5/2	4/1	SL	3msb	fr	a	
3c	C1	50-100	10YR 8/1	7/1	CL	m	fr	c	
	C2	100-155	10YR 8/1	7/2	SL	m	f	g	
2m	2Cg	155-162	2.5YR 5/2	4/2	CL	m	h	g	
	3C1	162-178	10YR 6/2	6/3	fS	sg	lo	a	
	4C2	178-200	10YR 3/3	2/1		m	fr	a	Interbeds of mud 1-3cm & sand <1cm; C14 sample.
1	5C3	200-240+	--------------	---	gS	sg	lo		

TABLE A1.32. DESCRIPTION OF WALKER CORE MU-17

Strat.	Horizon	Depth (cm)	Color Dry	Moist	Text.	Str.	Cons.	Bndy.	Remarks
Fill		0-17							
4s	A	17-48	7.5YR 4/2	3/2	SL	2msb	sh	g	
	BAw	48-70	7.5YR 4/2	3/2	SL	3msb	sh	c	
	Bw	70-97	7.5YR 4/4	3/4	SL	3msb	sh	c	
	Bk	97-155	7.5YR 7/2	4/2	SL	m	h	a	Finely divided carbonate.
3c	Akb	155-180	10YR 7/2	5/2	SL	3mpr& 2mab	h	c	Stage If; C14 sample.
	Cgb	180-275	10YR 8/2	---	SL	3fpr	vh	a	Common faint mottles of 2.5Y.
			2.5Y 7/2	---					
1	2Cb	275-335+	10YR 8/2	6/2	fS	sg	lo		Shell fragments common.

TABLE A1.33. DESCRIPTION OF WROE CORE MU-15

Strat.	Horizon	Depth (cm)	Color Dry	Moist	Text.	Str.	Cons.	Bndy.	Remarks
Fill		0-100							
4s	A	100-125	10YR 4/2	2/2	SiL	2msb	h	g	
	BAtk	125-180	10YR 5/3	---	SiI	2mpr& 3msb	h	c	Stage If; thin, cont. clay films.
			7.5YR 3/2	---					
	Btk1	180-195	7.5YR 4/3	3/2	SiL	3mpr& 3msb	h	c	Stage If; thin, cont. clay films.
	Btk2	195-205	7.5YR 7/2	5/4	SiL	3mpr& 3msb	vh	a	Stage IIf; thin, cont. clay films.
3c	Ab1	205-230	10YR 6/3	5/3	SiL	3fsb	vh	c	C14 sample 225-265cm.
	C1b1	230-280	7.5YR 6/3	5/3	SiL	2fsb	fr	c	
	C2b1	280-350	10YR 7/3	5/3	SiL	m	fr	c	Common snail shells; C14 sample 320-360cm.
	C3b1	350-395	10YR 7/3	5/2	L	m	fr	c	Common snail shells.
	C4b1	395-465	10YR 7/2	5/2	L	m	fr	a	Common snail shells; C14 sample 435-470cm.
2m	Ab2	465-485	10YR 7/3	5/2	L	m	sh	a	Noncalcareous; common snail shells & 1-5mm bodies of secondary silica.
1	2Cb2	485-500	10YR 7/2	6/2	cSgr	sg	lo		

APPENDIX 2. RADIOCARBON DATING THE VALLEY FILL

Vance T. Holliday and Herbert Haas

One of the most crucial aspects of this investigation of the late Quaternary record preserved in the valley fills of the Southern High Plains is establishing the geochronology. As of 1988, published, well-dated stratigraphic sequences had been established at the Clovis site, Lubbock Lake site, and Mustang Springs, along with one age each from Plainview and Gibson; ten unpublished radiocarbon ages also were available (Table A2.1). Between 1988 and 1992, 43 additional radiocarbon ages were secured from the draws (Table A2.1). Most of the ages were determined by the Southern Methodist University radiocarbon laboratory. A few others were determined by the Smithsonian Institution and by Beta Analytic, Inc., the latter for Brown et al. (1993).

All of the recently determined or earlier unpublished ages are from samples of organic-rich sediments and soil horizons. Preferred materials for radiocarbon dating such as charcoal or wood are very rare. Mollusc shell and bone are more abundant locally, but were avoided due to the uncertainties in interpreting ages from these materials. Two types of organic-rich sediment were encountered: (1) By volume most of these datable sediments are fine-grained (silty and clayey), homogeneous deposits up to 1 m thick which accumulated in slowly aggrading, probably marshy settings; (2) Less common were thin (<10 cm thick), clayey lenses interbedded with sand or diatomite and usually found in strata 2d or 2s. These layers probably were deposited relatively rapidly. Dating of dozens of these types of samples at Lubbock Lake, where archaeological data and radiocarbon ages on wood and charcoal also are available, show that dating these organic-rich samples yields reasonable, approximate ages of deposition (Holliday et al., 1983, 1985; Haas et al., 1986).

Organic matter in the A horizons of buried soils was incorporated into the surface of the soil parent material during a period of landscape stability and pedogenesis. A radiocarbon age from such "homogenized" horizons is the "mean residence time" of organic material accumulating in this zone, plus the time since burial by overlying sediments (Scharpenseel, 1971). Assuming they are then cut off from additional illuviation of organic matter or other forms of contamination, buried A horizons yield the maximum age of burial. Data from Lubbock Lake show that the age of burial, and therefore the age of the overlying sediments, can be reasonably approximated by collecting samples from the tops of these zones (Haas et al., 1986). Radiocarbon ages from the lower portions of a buried A horizon provide an intermediate age between the beginning and end of pedogenesis (Haas et al., 1986). Unless otherwise noted in Table A2.1, all A horizon samples were collected from the top 5 cm of the sampled strata or soils. Some of the resulting radiocarbon ages are considered unreliable (Table A2.1). This interpretation is based on stratigraphic considerations, other radiocarbon ages from the site (if available), and the regional chronostratigraphy.

Most radiocarbon ages listed in Table A2.1 were determined on the NaOH-soluble fraction of the sample (humic acid or "humates"). The humate fraction was dated by the SMU laboratory and by Beta Analytic. The two ages determined by the Smithsonian Institution are from the NaOH-insoluble fraction (humin or "residue"). The procedures used by the SMU and SI labs are outlined by Haas et al. (1986). Information on the procedures used by Beta are available directly from Beta Analytic, Inc. (Coral Gables, Florida).

All ages are based on a radiocarbon half-life of 5,568 years and most are corrected for $\delta^{13}C$ fractionation. The fractionation values are presented in Table 22 (for sites discussed in earlier section, "Paleontology, Paleobotany, and Stable Isotopes") and in Table A2.2. Radiocarbon ages were not dendrocalibrated.

**TABLE A2.1. RADIOCARBON AGES FROM LATE QUATERNARY VALLEY FILL,
SOUTHERN HIGH PLAINS***

Draw Site	Core	^{14}C Age[†] (yr B.P.)	Sample	Stratum	Comments[§]
Blackwater Draw					
Anderson	Bw-71	7,340 ± 180	SMU-2691	3s	Buried A horizon.
Basin #1	Bw-71	10,650 ± 140	SMU-2703	2m	OM-rich sediment, entire layer sampled.
Anderson Basin #2	Bw-58	5,755 ± 85**	SMU-2508	1s	Buried A horizon in sand.
Bailey	Bw-39	8,980 ± 70	SMU-2348	3c	Upper buried A horizon in marl.
	Bw-39	9,640 ± 240	SMU-2349	3c	Lower buried A horizon in marl.
BFI	____	5,125 ± 95	SI-5495	3s	Buried A horizon in calcareous sand; no fractionation correction.
	____	8,130 ± 50	SMU-1675	3c	Buried A horizon in marl.
Cannon	Bw-29	7,935 ± 210	SMU-2509	3c	Buried A horizon in marl.
Clovis	____	_____	_____	__	For discussion of previously published ages, see Haynes and Agogino, 1966; Haynes, 1975, 1995.
	Bw-59	21,140 ± 470	SMU-2533	A	Buried A horizon.
Davis	Bw-44	4,750 ± 80**	SMU-2536	3c	Buried A horizon in marl.
	Bw-45	10,910 ± 80	SMU-2343	2m	OM-rich mud; entire layer sampled.
Evans	Bw-17	340 ± 60	SMU-2510	5s1	Buried A horizon.
	Bw-17	3,540 ± 140	SMU-2245	3c	Buried A horizon.
Gibson	Bw-31	9,710 ± 80	SMU-2247	2d	OM-rich mud within diatomite; middle third sampled.
	----	9,920 ± 380	GX-1458	2d	Bone in diatomite, no fractionation correction; Marks Beach site; see Honea, 1980.
Halsell	Bw-27	8,530 ± 100	SMU-2248	3c	Buried A horizon in marl.
Jorde	----	16,590 + 510/-480	A-6901	A	Buried A horizon.
Lubbock Landfill	----	3,140 ± 100	Beta-43007	4s	Buried A horizon in loam.
	----	6,530 ± 80	Beta-43008	3si	Buried A horizon in loam.
	----	8,840 ± 120	Beta-43009	3c	Buried A horizon in marl.
	----	10,250 ± 100	Beta-57226	2m	OM-rich mud.
	----	10,540 ± 120	Beta-61962	2m	OM-rich mud.
Progress	Bw-42	9,930 ± 80	SMU-2344	2m	OM-rich mud, upper half sampled.
	Bw-42	11,330 ± 90	SMU-2345	2m	OM-rich mud, lower half sampled.
Tolk	Bw-36	10,090 ± 250	SMU2347	2m	OM-rich mud within sand; entire layer sampled.
Midland Draw					
Boone	Md-2	9,520 ± 260	SMU-2511	3c	Buried A horizon in marl.
Monahans Draw					
Midland	----	------	-----	---	For discussion of previously published ages, see Wendorf and Krieger, 1959.

**TABLE A2.1. RADIOCARBON AGES FROM LATE QUATERNARY VALLEY FILL,
SOUTHERN HIGH PLAINS*** (continued)

Draw Site	Core	¹⁴C Age[†] (yr B.P.)	Sample	Stratum	Comments[§]
Mustang Draw					
Wroe	Mu-15	5,720 ± 170	SMU-2503	3c	Buried A horizon in marl.
	Mu-15	8,345 ± 190	SMU-2506	3c	OM-rich sediments, middle 20 cm of layer.
	MU-15	10,120 ± 110	SMU-2507	2m	OM-rich sediments, middle 20 cm of layer.
Glendenning	Mu-3	9,470 ± 70	SMU-2249	3c	Buried A horizon in marl.
Mustang Springs	----	--------	------	---	For discussion of previously published ages, see Meltzer and Collins, 1987; Meltzer, 1991.
Walker	Mu-17	8,730 ± 80	SMU-2407	3c	Buried A horizon in marl.
Running Water Draw					
Edmonson	Rw-19	2,280 ± 50**	SMU-2412	3c	Buried A horizon in marl.
	Rw-19	9,760 ± 240	SMU-2413	2m/3c	OM-rich mud within sand; top of 2m and base of 3c.
Flagg	Rw-11	770 ± 50	SMU-2411	4s	Buried A horizon in loam.
	Rw-11	8,640 ± 100	SMU-2414	2d	OM-rich mud, upper half sampled.
	Rw-11	9,390 ± 120	SMU-2415	2m	OM-rich mud, lower half sampled.
Ned Houck	Rw-2	160 ± 50**	SMU-2409	2?	Buried A horizon with charcoal in loam.
	Rw-3	9,820 ± 50	SMU-2410	3c	Buried A horizon in marl.
Plainview	----	2,070 ± 130	SMU-1239	4s	Buried A horizon in loam.
	----	3,225 ± 65	SI-4820	5m	Bottom 10 cm of mud; NaOH-insoluble fraction; see Holliday, 1985b.
	----	3,880 ± 60	SMU-1234	4s	Buried A horizon in loam.
	----	6,770 ± 50	SMU-1349	3s	Buried A horizon in loam.
	----	8,860 ± 110	SMU-2341	2m	OM-rich mud within sand, entire layer sampled.
	----	10,940 ± 70	SMU-1359	1?	OM-rich mud, sampled from 10 cm zone starting 35 cm below top.
	----	11,970 ± 140	SMU-1376	1?	OM-rich mud, bottom 10 cm.
	----	--------	--------	---	For discussion of previously published ages, see Holliday, 1985b, 1990b; Johnson and Holliday, 1980; Holliday and Johnson, 1981; Speer, 1990.
Plains Paving	----	9,290 ± 90	SMU-1377	2m	OM-rich sediment, entire layer sampled.
Quincy Street	Rw-26	2,600 ± 40	SMU-2534	5m	OM-rich mud within sand and gravel.
Sunnyside	Rw-21	3,960 ± 80**	SMU-2535	4m	OM-rich sediment; base of stratum.
Seminole Draw					
Seminole-Rose	Se-13	16,310 ± 230	SMU-2342	B	Buried A horizon in sandy marl.
Sulphur Draw					
Brownfield	Su-16	9,730 ± 130	SMU-2408	3c	OM-rich sediment at base of stratum 4 and buried A horizon in stratum 3.
Sundown	Su-5	2,610 ± 50	SMU-2346	4s	Lower portion of overthickened A horizon of surface soil, formed in loamy eolian sediments

TABLE A2.1. RADIOCARBON AGES FROM LATE QUATERNARY VALLEY FILL,
SOUTHERN HIGH PLAINS* (continued)

Draw Site	Core	^{14}C Age† (yr B.P.)	Sample	Stratum	Comments§
Sulphur Springs Draw					
Lower Reach	----	--------	--------	---	For discussion of previously published ages, see Frederick, 1994.
Yellowhouse Draw					
Enochs	Yh-33	8,590 ± 80	SMU-2328	4s	Buried A horizon, base of stratum 4, above marl.
Lubbock Lake	----	--------	--------	---	For discussion of previously published ages, see Holliday et al., 1983, 1985; Haas et al., 1986.
Lupton	Yh-1	17,440 ± 840	SMU-2235	B	Buried A horizon in marl.
Payne	Yh-6	1,030 ± 50**	SMU-2244	3c	Buried A horizon in marl.

*Draws listed alphabetically; sites listed alphabetically for each draw and located on Figure 4.
†NaOH-soluble fraction unless otherwise noted; corrected for ^{13}C/^{12}C fractionation unless otherwise noted (Tables 22, A2.2); no dendrocalibration.
§OM = organic matter.
**Age considered unreliable.

REFERENCES CITED

AGI, 1982, AGI data sheets: Falls Church, Virginia, American Geological Institute, 157 p.

Anderson, D. M., 1962, The weevil genus *Smicronyx* in America north of Mexico (Coleoptera: Curculionidae): Proceedings, United States National Museum, v. 113, p. 324–327.

Antevs, E., 1935, The occurrence of flints and extinct animals in pluvial deposits near Clovis, New Mexico, Part III: Age of Clovis lake beds: Proceedings, Philadelphia Academy of Natural Sciences, v. 87, p. 304–311.

Antevs, E., 1949, Geology of the Clovis sites, *in* Wormington, H. M., Ancient man in North America: Denver Museum of Natural History, p. 185–192.

Antevs, E., 1954, Climate of New Mexico during the last glacio-pluvial: Journal of Geology, v. 62, p. 182–191.

Antevs, E., 1955, Geologic-climatic dating in the West: American Antiquity, v. 20, p. 317–335.

Ashworth, J. B., 1991, Water-level changes in the High Plains Aquifer of Texas, 1980–1990: Austin, Texas Water Development Board Hydrologic Atlas 1.

Ashworth, J. B., Christian, P., and Waterreus, T. C., 1991, Evaluation of groundwater resources in the Southern High Plains of Texas: Austin, Texas Water Development Board Report 330, 39 p.

Baerreis, D. A., and Bryson, R. A., 1965, Historical climatology and the Southern Plains: A preliminary statement: Bulletin of the Oklahoma Anthropological Society, v. 13, p. 69–75.

Baerreis, D. A., and Bryson, R. A., 1966, Dating the Panhandle Aspect cultures: Bulletin of the Oklahoma Anthropological Society, v. 14, p. 105–116.

Baker, C. L., 1915, Geology and underground waters of the northern Llano Estacado: The University of Texas at Austin Bulletin 57, 93 p.

Bark, L. D., 1978, History of American droughts, *in* Rosenberg, N. J., ed., North American droughts: Boulder, Colorado, Westview Press, p. 9–24.

Bartolino, J. R., 1991, Radon-222 in the groundwater surrounding the Anton Lake basin, Hockley County, Texas [Ph.D. thesis]: Lubbock, Texas Tech University, 205 p.

Bates, R. L., and Jackson, J. A. eds., 1980, Glossary of geology (second edition): Falls Church, Virginia, American Geological Institute, 751 p.

Becker, B., 1993, An 11,000-year German oak and pine dendrochronology for radiocarbon calibration: Radiocarbon, v. 35, p. 201–213.

Benninghoff, W. S., 1962, Calculation of pollen and spore density in sediments by addition of exotic pollen in known quantities: Pollen et Spores, v. 4, p. 332.

Bigham, J. M., Jaynes, W. F., and Allen, B. L., 1980, Pedogenic degradation of sepiolite and palygorskite on the Texas High Plains: Soil Science Society of America Journal, v. 44, p. 159–167.

Birkeland, P. W., 1984, Soils and geomorphology: New York, Oxford University Press, 372 p.

Blache, J., 1940, Le probleme des meandres encaisses et les rivieres Lorraines, III: Journal of Geomorphology, v. 3, p. 311–325.

Black, C. C., ed., 1974, History and prehistory of the Lubbock Lake site: West Texas Museum Association, The Museum Journal, v. 15, p. 1–160.

Blackstock, D. A., 1979, Soil survey of Lubbock County, Texas: Washington, D.C., U.S. Department of Agriculture, Soil Conservation Service, 105 p.

Blair, W. F., 1950, The biotic provinces of Texas: Texas Journal of Science, v. 2, p. 93–117.

Blum, M. D., and Valastro, S., Jr., 1992, Quaternary stratigraphy and geoarchaeology of the Colorado and Concho Rivers, west Texas: Geoarchaeology, v. 7, p. 419–448.

Blum, M. D., Abbot, J. T., and Valastro, S., Jr., 1992, Evolution of landscapes on the Double Mountain Fork of the Brazos River, west Texas: Implications for preservation and visibility of the archaeological record: Geoarchaeology, v. 7, p. 339–370.

Blum, M. D., Toomey, R. S., III, and Valastro, S., Jr., 1994, Fluvial response to late Quaternary climatic and environmental change, Edwards Plateau, Texas: Palaeogeography, Palaeoclimatology, Palaeoecology, v. 108, p. 1–21.

TABLE A2.2. δ¹³C VALUES FOR ORGANIC-RICH SEDIMENTS AND SOILS AT SITES ON RUNNING WATER, BLACKWATER, YELLOW-HOUSE, SULPHUR, SEMINOLE, MUSTANG, AND MIDLAND DRAWS*

Draw†	Site	Sample	Age (yr B.P.	Stratum	δ¹³C
Bw	Evans	SMU-2510	340 ± 60	5s	-17.1
	A.B. #1	SMU-2691	7,340 ± 180	3s	-20.5
	Cannon	SMU-2509	7,935 ± 210	3c	-15.9
	BFI	SMU-1675	8,130 ± 50	3c	-15.7
	Halsell	SMU-2248	8,530 ± 100	3c	-16.0
	Bailey	SMU-2348	8,980 ± 70	3c	-16.2
	Bailey	SMU-2349	9,640 ± 240	3c	-16.3
	Gibson	SMU-2247	9,710 ± 75	2d	-22.8
	Progress	SMU-2344	9,930 ± 80	2m	-16.6
	Tolk	SMU-2347	10,090 ± 250	2m	-19.2
	A.B. #1	SMU-2703	10,650 ± 140	2m	-23.6
	Davis	SMU-2343	10,910 ± 80	2m	-20.4
	Progress	SMU-2345	11,330 ± 90	2m	-16.6
	Jorde	A-6901	16,590 +510/-480	A	-19.4
	Clovis	SMU-2533	21,140 ± 470	A	-19.1
Md	Boone	SMU-2511	9,520 ± 260	3c	-15.0
Mu	Wroe	SMU-2503	5,720 ± 170	3c	-14.8
	Wroe	SMU-2506	8,345 ± 190	3c	-16.8
	Walker	SMU-2407	8,730 ± 80	3c	-19.1
	Glendenning	SMU-2249	9,470 ± 70	3c	-21.5
	Wroe	SMU-2507	10,120 ± 110	2m	-20.5
Se	Seminole-Rose	SMU-2342	16,310 ± 225	B	-15.4
Su	Sundown	SMU-2346	2,610 ± 50	4s	-14.2
	Brownfield	SMU-2408	9,730 ± 130	3c	-17.2
Rw	Flagg	SMU-2411	770 ± 50	4s	-15.9
	Quincy St.	SMU-2534	2,600 ± 40	4m	-15.5
	Flagg	SMU-2414	8,640 ± 100	2d	-22.7
	Plainview	SMU-2341	8,860 ± 110	2m	-16.4
	Plains Paving	SMU-1377	9,290 ± 90	2m	-20.5
	Flagg	SMU-2415	9,390 ± 120	2m	-21.3
	Edmonson	SMU-2413	9,760 ± 240	2m	-22.1
	Houck	SMU-2410	9,820 ± 50	3c	-15.9
Yh	Enochs	SMU-2328	8,590 ± 80	4s	-16.0
	Lupton	SMU-2235	17,440 ± 840	B	-16.8

*For Lubbock Lake, Lubbock Landfill, Mustang Springs, and Plainview, see Table 22.

†Bw = Blackwater; Md = Midland; Mu = Mustang; Se = Seminole; Su = Sulphur; Rw = Running Water; Yh = Yellowhouse.

Boldurian, A., 1990, Lithic technology at the Mitchell locality of Blackwater Draw: Plains Anthropologist Memoir, v. 24, 115 p.

Bolton, H. E., 1990, Coronado, knight of pueblos and plains (reprint of 1949 edition): Albuquerque, The University of New Mexico Press, 491 p.

Borchert, J. R., 1971, The dust bowl in the 1970's: Annals of the Association of American Geographers, v. 61, p. 1–21.

Bowen, R., 1991, Isotopes and climates: New York, Elsevier Applied Science, 483 p.

Bozarth, S. R., 1986, Phytoliths, in Blakeslee, D. J., Blasing, R., and Garcia, H., authors, Along the Pawnee Trail, cultural resource survey and testing, Wilson Lake, Kansas: For the Kansas City District U.S. Corps of Engineers, p. 86–101.

Bozarth, S. R., 1988, Preliminary opal phytolith analysis of modern analogs from parklands, mixed forest, and selected conifer stands in Prince Albert National Park, Saskatchewan: Current Research in the Pleistocene, v. 5, p. 45–46.

Bozarth, S. R., 1989, Presence/absence opal phytolith analysis of late Quaternary sediment from the Southern High Plains: Manuscript on file at the University of Wisconsin—Madison, Department of Geography, 7 p.

Bozarth, S. R., 1991, Extracting pollen and phytoliths from archaeological sediment, in The Roosevelt rural sites study—Laboratory manual: Tucson, Arizona, Statistical Research, submitted to the U.S. Department of Interior, Bureau of Reclamation, Arizona Projects Office, p. VII6–VII7.

Bozarth, S. R., 1992, Classification of opal phytoliths formed in selected dicotyledons native to the Great Plains, in Rapp, G., Jr., and Mulholland, S., eds., Phytolith systematics—emerging issues: New York, Plenum Press, p. 193–214.

Bozarth, S. R., 1994, Biosilicate assemblages of Boreal forests and Aspen parklands, in Pearsall, D., and Piperno, D., eds., Current research in phytolith analysis: Applications in archaeology and paleoecology: Philadelphia, University of Pennsylvania, MASCA (Museum Applied Science Center for Archaeology) series, p. 95–105.

Brown, D. O., Holliday, V. T., Anthony, D., Cargill, D., and Arthur, J., 1993, Archeological survey of a proposed landfill, City of Lubbock, Lubbock County, Texas: Austin, Hicks and Co., Inc., Archeology Series 22, 120 p.

Brune, G., 1981, Springs of Texas: Fort Worth, Branch-Smith, Inc., 566 p.

Bryan, F., 1938, A review of the geology of the Clovis finds reported by Howard and Cotter: American Antiquity, v. 4, p. 113–130.

Bryant, V. M., Jr., and Schoenwetter, J., 1987, Pollen records from the Lubbock Lake site, in Johnson, E., ed., Lubbock Lake: Late Quaternary studies on the Southern High Plains: College Station, Texas A&M University Press, p. 36–40.

Bryant, V. M., Jr., and Hall, S. A., 1993, Archaeological palynology in the United States: A critique: American Antiquity, v. 58, p. 277–286.

Bull, W. B., 1991, Geomorphic responses to climatic change: New York, Oxford University Press, 326 p.

Burch, J. B., 1962, How to know the eastern land snails: Dubuque, Iowa, W. C. Brown, Co., 214 p.

Burch, J. B., 1972, Freshwater sphaeriacean clams (Mollusca: Pelecypoda) of North America, Biota of freshwater ecosystems, Identification manual 3: in Washington, D.C., Environmental Protection Agency, 31 p.

Burch, J. B., 1975, Freshwater unionacean clams (Mollusca: Pelecypoda) of North America: Hamburg, Michigan, Malacological Publications, 204 p.

Burch, J. B., 1982, Freshwater snails (Molluscs: Gastropoda) of North America: Cincinnati, U.S. Environmental Protection Agency, EPA-600/3-82-026, p. 1–294.

Bureau of Immigration, 1903, Roosevelt County, New Mexico (pamphlet): Bureau of Immigration of the Territory of New Mexico, 29 p.

Campbell, C. A., Paul, E. A., Rennie, D. A., and McCallum, K. J., 1967, Factors affecting the accuracy of the carbon-dating method in soil humus studies: Soil Science, v. 104, no. 2, p. 81–85.

Caran, S. C., 1991, Cenozoic stratigraphy, southern Great Plains area, in Morrison, R. B., ed., Quaternary nonglacial geology: Conterminous U.S.: Boulder, Geological Society of America Centennial Volume K-2, Plate 5.

Caran, S. C., and Baumgardner, R. W., Jr., 1990, Quaternary stratigraphy and paleoenvironments of the Texas Rolling Plains: Geological Society of America Bulletin, v. 102, p. 768–785.

Caran, S. C., and Baumgardner, R. W., Jr., 1991, Quaternary geology of the Texas Rolling Plains, *in* Morrison R. B., ed., Quaternary nonglacial geology: Conterminous U.S.: Geological Society of America Centennial Volume K-2, p. 492–497.

Cerling, T. E., 1984, The stable isotope composition of modern soil carbonate and its relationship to climate: Earth and Planetary Science Letters, v. 71, p. 229–240.

Clarke, W. T., Jr., 1938, The occurrence of flints and extinct animals in pluvial deposits near Clovis, New Mexico, Part VII: The Pleistocene mollusks from the Clovis gravel pit and vicinity: Proceedings, Philadelphia Academy of Natural Sciences, v. 90, p. 119–121.

COHMAP Members, 1988, Climatic changes of the last 18,000 years: Observations and model simulations: Science, v. 241, p. 1043–1052.

Coope, G. R., 1970, Interpretations of Quaternary insect fossils: Annual Review of Entomology, v. 15, p. 97–120.

Cotter, J. L., 1937, The occurrence of flints and extinct animals in pluvial deposits near Clovis, New Mexico, Part IV: Report on excavation at the gravel pit, 1936: Proceedings, Philadelphia Academy of Natural Sciences, v. 90, p. 2–16.

Cotter, J. L., 1938, The occurrence of flints and extinct animals in pluvial deposits near Clovis, New Mexico, Part VI: Report on field season of 1937: Proceedings, Philadelphia Academy of Natural Sciences, v. 90, p. 113–117.

Cronin, J. G., 1964, A summary of the occurrence and development of ground water in the Southern High Plains of Texas: U.S. Geological Survey Water-Supply Paper 1693, 88 p.

Cronin, J. G., 1969, Ground water in the Ogallala Formation in the Southern High Plains of Texas and New Mexico: U.S. Geological Survey Atlas HA-330, Sheet 1, 1:500,000.

Dalquest, W. W., 1986, Vertebrate fossils from a strath terrace of Quitaque Creek, Motley County, Texas, *in* Gustavson, T. C., ed., Geomorphology and Quaternary stratigraphy of the Rolling Plains, Texas Panhandle: Austin, The University of Texas at Austin Bureau of Economic Geology, Guidebook 22, p. 58–59.

Dalquest, W. W., and Schultz, G. E., 1992, Ice Age mammals of northwestern Texas: Wichita Falls, Texas, Midwestern State University Press, 309 p.

Day, P. R., 1965, Particle fractionation and particle-size analysis, *in* Black, C. A., ed., Methods of soil analysis, Part I: Madison, Wisconsin, Soil Science Society of America, Agronomy Series 9, p. 545–567.

DeDeckker, P., and Forester, R. M., 1988, The use of Ostracods to reconstruct continental palaeoenvironmental records, *in* DeDeckker, P., Colin, J. P., and Peypouquet, J. P., eds. Ostracode in the earth sciences: Amsterdam, Elsevier, p. 175–199.

Drake, R. J., 1975, Fossil nonmarine molluscs of the 1961–63 Llano Estacado paleoecology study, *in* Wendorf, F., and Hester, J. J., eds., Late Pleistocene environments of the Southern High Plains: Taos, Publications of the Fort Burgwin Research Center, v. 9, p. 201–245.

Eifler, G. K., Jr., and Reeves, C. C., Jr., 1976, Hobbs sheet: The University of Texas at Austin Bureau of Economic Geology, Geologic Atlas of Texas, scale 1:250,000.

Elias, S. A., 1987, Colorado ground beetles (Coleoptera: Carabidae) from the Rotger collection, University of Colorado Museum: Great Basin Naturalist, v. 47, p. 631–637.

Elias, S. A., 1994, Quaternary insects and their environments: Washington, D.C., Smithsonian Institution Press, 284 p.

Elias, S. A., and Johnson, E., 1988, Pilot study of fossil beetles at the Lubbock Lake Landmark: Current Research in the Pleistocene, v. 5, p. 57–59.

Evans, G. L., 1949, Upper Cenozoic of the High Plains, *in* Cenozoic Geology of the Llano Estacado and Rio Grande Valley: Lubbock, West Texas Geological Society Guidebook, v. 2, p. 1–22.

Evans, G. L., 1951, Prehistoric wells in eastern New Mexico: American Antiquity, v. 17, p. 1–8.

Evans, G. L., and Brand, J. P., 1956, Eastern Llano Estacado and adjoining Osage Plain: Lubbock, West Texas Geological Society and Lubbock Geological Society, 1956 Spring Field Trip Guidebook, 102 p.

Evans, G. L., and Meade, G. E., 1945, Quaternary of the Texas High Plains: The University of Texas at Austin Publication 4401, p. 485–507.

Fallin, J. A., Nordstrom, P., and Ashworth, J. B., 1987, 1986 water level rises and recharge patterns in the Southern High Plains aquifer of Texas, *in* Symposium on the Quaternary Blackwater Draw and Tertiary Ogallala formations, Program with Abstracts: The University of Texas at Austin Bureau of Economic Geology, p. 10.

Fenneman, N. M., 1931, Physiography of western United States: New York, McGraw-Hill Book Company, 534 p.

Ferring, C. R., 1990, Late Quaternary geology and geoarchaeology of the upper Trinity River drainage basin, Texas, Geological Society of America, Field Trip 11 Guidebook,: Dallas, Dallas Geological Society, 79 p.

Ferring, C. R., 1994, The role of geoarchaeology in Paleoindian research, in Bonnichsen, R., and Steele, D. G., eds., Method and theory for investigating the peopling of the Americas: Corvallis, Oregon, Center for the Study of the First Americans, p. 57–72.

Fiedler, A. G., and Nye, S. S., 1933, Geology and ground-water resources of the Roswell artesian basin, New Mexico: U.S. Geological Survey Water-Supply Paper 639, 372 p.

Finch, W. I., and Wright, J. C., 1970, Linear features and ground-water distribution in the Ogallala Formation of the Southern High Plains, *in* Mattox, R. B., and Miller, W. D., eds. Ogallala Aquifer Symposium: Lubbock, Texas Tech University, ICASALS (International Center for Arid and Semi-Arid Land Studies), p. 49–57.

Finley, R. J., and Gustavson, T. C., 1983, Geomorphic effects of a 10-year storm on a small drainage basin in the Texas Panhandle: Earth Surface Processes and Landforms, v. 8, p. 63–77.

Foged, N., 1981, Diatoms in Alaska: Vaduz, J. Cramer, 317 p.

Foged, N., 1984, The diatom flora in springs in Jutland, Denmark (Springs III): Bibliotheca Diatomologica Band 4: Vaduz, J. Cramer, p. 1–119.

Forester, R. M., 1987, Late Quaternary paleoclimate records from lacustrine ostracodes, *in* Ruddiman, W. F., and Wright, H. E., Jr. eds., North America and adjacent oceans during the last deglaciation: Boulder, Colorado, Geological Society of America Centennial Volume K-3, p. 261–276.

Forester, R. M., 1991a, Ostracode assemblages from springs in the western United States: Implications for paleohydrology, *in* Williams, D. D., and Danks, H. V., eds. Arthropods of springs, with particular reference to Canada: Memoirs of the Entomological Society of Canada, v. 155, p. 181–201.

Forester, R. M., 1991b, Pliocene climate history of the western United States derived from lacustrine ostracodes: Quaternary Science Reviews, v. 10, p. 133–146.

Forester, R. M., Colman, S. M., Reynolds, R. L., and Keigwin, L. D., 1994, Lake Michigan's late Quaternary limnological and climate history from ostracode, oxygen isotope, and magnetic susceptibility: Journal of Great Lakes Research, v. 20, p. 93–107.

Frederick, C. D., 1993a, Geomorphic investigations, *in* Quigg, J. M., Frederick, C. D., and Lintz, C., authors, Sulphur Springs Draw: Archaeological and geomorphological investigations at Red Lake Dam axis, borrow area, and spillway, Martin County, Texas: Austin, Mariah Associates, Inc., Mariah Technical Report 873, p. 36–49.

Frederick, C. D., 1993b, Geomorphology, *in* Quigg, J. M., Lintz, C., Oglesby, F. M., Earls, A. C., Frederick, C. D., Trierweiler, W. N., Owsley, D., and Kibler, K. W., authors, Historic and prehistoric data recovery at Palo Duro Reservoir, Hansford County, Texas: Austin, Mariah and Associates, Inc., Mariah Technical Report 485, p. 75–116.

Frederick, C. D., 1994, Late Quaternary geology of the Sulphur Draw Reservoir, *in* Quigg, J. M., Frederick, C. D., and Lintz, C., authors, Sulphur Springs Draw: Geoarchaeological and archaeological investigations at Sulphur Draw Reservoir, Martin County, Texas: Austin, Mariah Associates, Inc., Mariah Technical Report 776, p. 29–82.

Fredlund, G. G., Johnson, W. C., and Dort, W., Jr. 1985, A preliminary analysis of opal phytoliths from the Eustis Ash Pit, Frontier County, Nebraska, *in* Dort, W., ed., Institute for Tertiary Quaternary Studies, TER-QUA Symposium Series, Volume 1: Lincoln, Nebraska Academy of Sciences, Inc., p. 147–162.

Frye, J. C., and Leonard, A. B., 1964, Relation of Ogallala Formation to the Southern High Plains in Texas: The University of Texas at Austin Bureau of Economic Geology, Report of Investigations 51, 25 p.

Frye, J. C., and Leonard, A. B., 1968, Late-Pleistocene Lake Lomax in western Texas, *in* Morrison, R. B., and Wright, H. E., Jr., eds., Means of correlation of Quaternary successions: Salt Lake City, University of Utah Press, p. 519–534.

Gable, D. J., and Hatton, T., 1983, Maps of vertical crustal movements in the conterminous United States over the last 10 million years: U.S. Geological Survey Map I-1315, Map B, 1:5,000,000.

Gardner, T. W., 1975, The history of part of the Colorado River and its tributaries: an experimental study, *in* Fasset, J. E., ed, Four Corners Geological Society, 8th Field Conference, Canyonlands, Utah Guidebook, Durango, Colorado, Four Corners Geological Society, p. 87–95.

Gasse, F., 1986, East African diatom taxonomy and ecological distribution, Bibliotheca Diatomologica Band 11: Berlin, J. Cramer, 201 p.

Gasse, F., 1987, Diatoms for reconstructing paleoenvironments and paleohydrology in tropical semi-arid zones, example of some lakes from Niger since 12,000 BP: Hydrobiologia, v. 154, p. 127–163.

Gazdar, M. N., 1981, Tertiary and Quaternary drainage of the Southern High Plains [Ph.D. thesis]: Lubbock, Texas Tech University, 110 p.

Geis, J. W., 1973, Biogenic silica in selected species of deciduous angiosperms: Soil Science, v. 116, p. 113–130.

Gidaspow, T., 1959, North American caterpillar hunters of the genera *Calosoma* and *Callisthenes*: American Museum of Natural History Bulletin, v. 116, p. 227–343.

Gile, L. H., 1979, Holocene soils in eolian sediments of Bailey County, Texas: Soil Science Society of America Journal, v. 43, p. 994–1003.

Gile, L. H., Peterson, F. F., and Grossman, R. B., 1966, Morphological and genetic sequences of carbonate accumulation in desert soils: Soil Science, v. 101, p. 347–360.

Gordon, R. D., and Cartwright, O. L., 1988, North America representatives of Aegialiini (Coleoptera: Scarabaeidae: Aphodiinae): Smithsonian Contributions to Zoology, no. 461, 37 p.

Graham, R. W., 1987, Late Quaternary mammalian faunas and paleoenvironments of the southwestern plains of the United States, *in* Graham, R. W., Semken, H. A., Jr., and Graham, M. A., eds., Late Quaternary mammalian biogeography and environments of the Great Plains and prairies: Illinois State Museum Scientific Papers, v. 22, p. 24–86.

Green, F. E., 1962a, The Lubbock Reservoir site: West Texas Museum Association, The Museum Journal, v. 6, p. 83–123.

Green, F. E., 1962b, Additional notes on prehistoric wells at the Clovis site: American Antiquity, v. 28, p. 230–234.

Green, F. E., 1963, The Clovis blades: An important addition to the Llano Complex: American Antiquity, v. 29, p. 145–165.

Green, F. E., 1992, Comments on the report of a worked mammoth tusk from the Clovis site: American Antiquity, v. 57, p. 331–337.

Guffee, E. J., 1979, The Plainview site: relocation and archeological investigation of a late Paleo-Indian kill site in Hale County, Texas: Plainview, Texas, Llano Estacado Museum, Archeological Research Laboratory, 64 p.

Gustavson, T. C., and Finley, R. C., 1985, Late Cenozoic geomorphic evolution of the Texas Panhandle and northeastern New Mexico: The University of Texas at Austin Bureau of Economic Geology, Report of Investigations 148, 42 p.

Gustavson, T. C., and Holliday, V. T., 1988, Depositional and pedogenic characteristics of the Quaternary Blackwater Draw and Tertiary Ogallala Formations, Texas Panhandle and eastern New Mexico: The University of Texas at Austin Bureau of Economic Geology, Open-File Report OF-

WTWI-1985-23 (revision 1), 112 p.

Gustavson, T. C., and Winkler, D. A., 1988, Depositional facies of the Miocene-Pliocene Ogallala Formation, northwestern Texas and eastern New Mexico: Geology, v. 16, p. 203–206.

Gustavson, T. C., Finley, R. J., and McGillis, K. A., 1980, Regional dissolution of Permian salt in the Anadarko, Dalhart, and Palo Duro basins of the Texas Panhandle: The University of Texas at Austin Bureau of Economic Geology, Report of Investigation 106, 40 p.

Gustavson, T. C., Baumgardner, R. W., Jr., Caran, S. C., Holliday, V. T., Mehnert, H. H., O'Neill, J. M., and Reeves, C. C., Jr. 1990, Quaternary geology of the southern Great Plains and an adjacent segment of the Rolling Plains, *in* Morrison, R. B., ed., Quaternary nonglacial geology: Conterminous U.S.: Geological Society of America Centennial Volume K-2, p. 477–501.

Gutentag, E. D., Heimes, F. J., Krothe, N. C., Luckey, R. R., and Weeks, J. B., 1984, Geohydrology of the High Plains aquifer in parts of Colorado, Kansas, Nebraska, New Mexico, Oklahoma, South Dakota, Texas, and Wyoming: U.S. Geological Survey Professional Paper 1400-B, 63 p.

Haas, H. H., Holliday, V. T., and Stuckenrath, R., 1986, Dating of Holocene stratigraphy with soluble and insoluble organic fractions at the Lubbock Lake archaeological site, Texas: an ideal case study: Radiocarbon, v. 28, p. 473–485.

Hafsten, U., 1961, Pleistocene development of vegetation and climate in the Southern High Plains as evidenced by pollen analysis, *in* Wendorf, F., assembler, Paleoecology of the Llano Estacado: Santa Fe, Fort Burgwin Research Center Publication 1, The Museum of New Mexico Press, p. 59–91.

Hafsten, U., 1964, A standard pollen diagram for the southern High Plains, USA, covering the period back to the early Wisconsin glaciation: International Quaternary Congress, 6th, Warsaw, 1961 Report 2, p. 407–420.

Hageman, B. P., 1972, Reports of the International Quaternary Association Subcommission on the study of the Holocene, Bulletin 6, 6 p.

Haley, J. E., 1953, The XIT Ranch of Texas and the early days of the Llano Estacado: Norman, University of Oklahoma Press, 258 p.

Hall, S. A., 1982, Late Holocene paleoecology of the Southern Plains: Quaternary Research, v. 17, p. 391–407.

Hall, S. A., and Lintz, C., 1984, Buried trees, water table fluctuations, and 3000 years of changing climate in west-central Oklahoma: Quaternary Research, v. 22, p. 129–133.

Hall, S. A., and Valastro, S., 1995, Grassland vegetation in the Southern Great Plains during the last glacial maximum: Quaternary Research, in press.

Hammond, A. P., Goh, K. M., Tonkin, P. J., and Manning, M. R., 1991, Chemical pretreatments for improving the radiocarbon dates of peats and organic silts in a gley podzol environment: Grahams terrace, North Westland: New Zealand Journal of Geology and Geophysics, v. 34, p. 191–194.

Hansen, J., Fung, I., Lacis, A., Rind, D., Lebedeff, S., Ruedy, R., and Russell, G., 1988, Global climate changes as forecast by Goddard Institute for Space Studies three dimensional model: Journal of Geophysical Research, v. 93(D8), p. 9341–9364.

Harbour, J., 1975, General stratigraphy, *in* Wendorf, F., and Hester, J. J., eds., Late Pleistocene environments of the Southern High Plains: Taos, Publications of the Fort Burgwin Research Center, v. 9, p. 33–55.

Harden, D. P., 1990, Controlling factors in the distribution and development of incised meanders in the central Colorado Plateau: Geological Society of America Bulletin, v. 102, p. 233–242.

Hawley, J. W., Backman, G. O., and Manley, K., 1976, Quaternary stratigraphy in the Basin and Range and Great Basin Provinces, New Mexico and western Texas, *in* Mahaney, W. C., ed., Quaternary stratigraphy of North America: Stroudsburg, Pennsylvania, Dowden, Hutchinson and Ross, Inc., p. 235–274.

Haynes, C. V., Jr., 1968, Geochronology of late-Quaternary alluvium, *in* Morrison, R. B., and Wright, H. E., eds., Means of correlation of Quaternary successions: Salt Lake City, University of Utah Press, p. 591–631.

Haynes, C. V., Jr., 1975, Pleistocene and Recent stratigraphy, *in* Wendorf, F.,

and Hester, J. J., eds., Late Pleistocene environments of the Southern High Plains: Taos, Publication of the Fort Burgwin Research Center, v. 9, p. 57–96.

Haynes, C. V., Jr., 1991, Geoarchaeological and paleohydrological evidence for a Clovis age drought in North America and its bearing on extinction: Quaternary Research, v. 35, p. 438–450.

Haynes, C. V., Jr., 1993, Clovis-Folsom geochronology and climatic change, *in* Soffer, O., and Praslov, N. D., eds., From Kostenki to Clovis: Upper Paleolithic–Paleo-Indian adaptations: New York, Plenum Press, p. 219–236.

Haynes, C. V., Jr., 1995, Geochronology of paleoenvironmental change, Clovis type site, Blackwater Draw, New Mexico: Geoarchaeology, in press.

Haynes, C. V., Jr., and Agogino, G. A., 1966, Prehistoric springs and geochronology of the Clovis site, New Mexico: American Antiquity, v. 31, p. 812–821.

Haynes, C. V., Jr., Saunders, J. J., Stanford, D., and Agogino, G. A., 1992, F. E. Green's comments on the Clovis site: Reply: American Antiquity, v. 57, p. 338–344.

Hester, J. J., 1962, A Folsom lithic complex from the Elida site, Roosevelt County, N.M.: El Palacio, v. 69, p. 92–113.

Hester, J. J., compiler, 1972, Blackwater Locality no. 1: a stratified Early Man site in eastern New Mexico: Taos, Publication of the Fort Burgwin Research Center, v. 8, 239 p.

Hester, J. J., 1975, The sites, *in* Wendorf, F., and Hester, J. J., eds., Late Pleistocene environments of the Southern High Plains: Taos, Publication of the Fort Burgwin Research Center, v. 9, p. 13–32.

Hofman, J. L., Brooks, R. L., Hays, J. S., Owsley, D., Jantz, R. L., Marks, M. K., and Manhein, M. H., 1989, From Clovis to Comanchero: Archeological overview of the Southern Great Plains: Arkansas Archeological Survey Research Series 35, 286 p.

Hohn, M. H., 1975, The diatoms, *in* Wendorf, F., and Hester, J. J., eds., Late Pleistocene environments of the Southern High Plains: Taos, Publication of the Fort Burgwin Research Center, v. 9, p. 197–200.

Hohn, M. H., and Hellerman, J., 1961, The diatoms, *in* Wendorf, F., assembler, Paleoecology of the Llano Estacado: Santa Fe, The Museum of New Mexico Press, Burgwin Research Center Publication 1, p. 98–104.

Holden, W. C., 1959, Indians, Spaniards and Anglos, *in* Graves, L. L., ed., A history of Lubbock: The Museum Journal, West Texas Museum Association, v. 8, p. 17–44.

Holden, W. C., 1974, Historical background of the Lubbock Lake site, *in* Black, C. C., ed., History and prehistory of the Lubbock Lake site: The Museum Journal, West Texas Museum Association, v. 14, p. 11–14.

Holliday, V. T., 1982, Morphological and chemical trends in Holocene soils at the Lubbock Lake archaeological site, Texas [Ph.D. thesis]: Boulder, University of Colorado, 284 p.

Holliday, V. T., ed., 1983, Guidebook to the Central Llano Estacado, Friends of the Pleistocene South-Central Cell Field Trip: Lubbock, Texas Tech University, ICASALS (International Center for Arid and Semi-Arid Land Studies) and The Museum, 165 p.

Holliday, V. T., 1985a, Holocene soil-geomorphological relations in a semi-arid environment: the Southern High Plains of Texas, *in* Boardman, J., ed, Soils and Quaternary landscape evolution: New York, John Wiley and Sons, p. 325–357.

Holliday, V. T., 1985b, New data on the stratigraphy and pedology of the Clovis and Plainview sites, Southern High Plains: Quaternary Research, v. 23, p. 388–402.

Holliday, V. T., 1985c, Archaeological geology of the Lubbock Lake site, Southern High Plains of Texas: Geological Society of America Bulletin, v. 96, p. 1483–1492.

Holliday, V. T., 1985d, Morphology of late Holocene soils at the Lubbock Lake site, Texas: Soil Science Society of America Journal, v. 49, p. 938–946.

Holliday, V. T., 1985e, Early Holocene soils at the Lubbock Lake archaeological site, Texas: Catena, v. 12, p. 61–78.

Holliday, V. T., 1987, Re-examination of late-Pleistocene boreal forest recon-

structions for the Southern High Plains: Quaternary Research, v. 28, p. 238–244.

Holliday, V. T., 1988a, Genesis of late Holocene soils at the Lubbock Lake archaeological site, Texas: Annals of the Association of American Geographers, v. 78, p. 594–610.

Holliday, V. T., 1988b, Mt. Blanco revisited: soil-geomorphic implications for ages of the upper Cenozoic Blanco and Blackwater Draw Formations: Geology, v. 16, p. 505–508.

Holliday, V. T., 1989a, Middle Holocene drought on the Southern High Plains: Quaternary Research, v. 31, p. 74–82.

Holliday, V. T., 1989b, The Blackwater Draw Formation (Quaternary): a 1.4-plus m.y. record of eolian sedimentation and soil formation on the Southern High Plains: Geological Society of America Bulletin, v. 101, p. 1598–1607.

Holliday, V. T., 1990a, Soils and landscape evolution of eolian plains: the Southern High Plains of Texas and New Mexico, *in* Knuepfer, P. L. K., and McFadden, L. D., eds., Soils and landscape evolution: Geomorphology, v. 3, p. 489–515.

Holliday, V. T., 1990b, Investigations of the Plainview site and middle Running Water Draw, *in* Holliday, V. T., and Johnson, E., eds., Guidebook to the Quaternary history of the Llano Estacado: Lubbock, Texas Tech University, Lubbock Lake Landmark Quaternary Research Series, v. 2, p. 93–104.

Holliday, V. T., 1995, Late Quaternary stratigraphy of the Southern High Plains, *in* Johnson, E., ed., Ancient peoples and landscapes: Lubbock, Museum of Texas Tech University, p. 289–313.

Holliday, V. T., and Johnson, E., 1981, An update on the Plainview occupation at the Lubbock Lake site: Plains Anthropologist, v. 26, p. 251–253.

Holliday, V. T., Johnson, E., Haas, H., and Stuckenrath, R., 1983, Radiocarbon ages from the Lubbock Lake site, 1950–1980: Framework for cultural and ecological change on the Southern High Plains: Plains Anthropologist, v. 28, p. 165–182.

Holliday, V. T., Johnson, E., Haas, H., and Stuckenrath, R., 1985, Radiocarbon ages from the Lubbock Lake site, 1981-1984: Plains Anthropologist, v. 30, p. 277–291.

Holliday, V. T., Haynes, C. V., Jr., Hofman, J. L., and Meltzer, D. J., 1994, Geoarchaeology and geochronology of the Miami (Clovis) site, Southern High Plains of Texas: Quaternary Research, v. 41, p. 234–244.

Honea, K., 1980, the Marks Beach stratified Paleoindian site, Lamb County, Texas: Bulletin of the Texas Archeological Society, v. 51, p. 243–269.

Howard, E. B., 1935a, The occurrence of flints and extinct animals in pluvial deposits near Clovis, New Mexico, Part I: Introduction: Proceedings of the Academy of Natural Sciences of Philadelphia, v. 87, p. 299–303.

Howard, E. B., 1935b, Evidence of Early Man in North America: The Museum Journal, University of Pennsylvania Museum, v. 24, p.61–171.

Howden, H. F., and Cartwright, O. L., 1963, Scarab beetles of the genus *Onthophagus* Latreille north of Mexico (Coleoptera: Scarabaeidae): Proceedings, United States National Museum, v. 114, v. 1–135.

Hughes, J. T., and Guffee, E., 1976, Summary report on backhoe testing in the lower Running Water Draw watershed, Hale and Castro Counties, Texas: Temple, Texas, Report submitted to the Soil Conservation Service, 3 p.

Humphrey, J. D., and Ferring, C. R., 1994, Stable isotopic evidence for latest Pleistocene and Holocene climatic change in north-central Texas: Quaternary Research, v. 41, p. 200–213.

Hunt, C. B., 1974, Natural regions of the United States and Canada: San Francisco, W. H. Freeman and Co., 725 p.

Hustedt, F., 1930, Bacillariophyta, Pascher, Susswasserflora von Mitteleuropa: Heft, v. 10, p. 1–466.

Hustedt, F., 1927-1966, Die Kieselalgen Deutschlands, Osterreichs und der Schweiz mit Berucksichtigung der ubrigen Lander Europas sowie der angrenzenden Meeresgebiete: I, II, III: Rabenhorst Kryptogamenflora, Band VII, Teil 1, p. 1–920; Teil 2, p. 1–845; Teil 3, p. 1–816.

Jackson, M. L., 1969, Soil chemical analysis—advanced course (second edition): Madison, Wisconsin, published by the author, 895 p.

Janitzky, P., 1986a, Particle-size analysis, in Singer, M. J., and Janitzky, P., eds., Field and laboratory procedures used in a soil chronosequence study: U.S. Geological Survey Bulletin 1648, p. 11–16.

Janitzky, P., 1986b, Organic carbon (Walkley-Black method), in Singer, M. J., and Janitzky, P., eds., Field and laboratory procedures used in a soil chronosequence study: U.S. Geological Survey Bulletin 1648, p. 34-36.

Janitzky, P., 1986c, Determination of soil pH, in Singer, M. J., and Janitzky, P., eds., Field and laboratory procedures used in a soil chronosequence study: U.S. Geological Survey Bulletin 1648, p. 19–21.

Johnson, C. A., 1974, Geologic investigations at the Lubbock Lake site, in Black, C. C., ed., History and prehistory of the Lubbock Lake site: West Texas Museum Association, The Museum Journal, v. 15, p. 79–105.

Johnson, C. A., and Stafford, T. W., 1976, Report of 1975 of the archaeological survey investigation of the Canyon Lakes Project, Yellowhouse Canyon, Lubbock, Texas: Report submitted to the Texas Antiquities Committee and City of Lubbock, 25 p.

Johnson, E., 1986, Late Pleistocene and early Holocene paleoenvironments on the Southern High Plains (USA); Geographie Physique et Quaternaire, v. 40, p. 249–261.

Johnson, E., 1987a, Paleoenvironmental overview, in Johnson, E., ed., Lubbock Lake: Late Quaternary studies on the Southern High Plains: College Station, Texas A&M University Press, p. 90–99.

Johnson, E., ed., 1987b, Lubbock Lake: Late Quaternary studies on the Southern High Plains: College Station, Texas University Press, 179 p.

Johnson, E., 1987c, Vertebrate remains, in Johnson, E., ed., Lubbock Lake: Late Quaternary studies on the Southern High Plains: College Station, Texas A&M University Press, P. 49–89.

Johnson, E., 1987d, Cultural activities and interactions, in Johnson, E., ed., Lubbock Lake: Late Quaternary studies on the Southern High Plains: College Station, Texas A&M University Press, p. 120–158.

Johnson, E., ed., 1994, Archaeological survey along the Mobil ESTE CO_2 pipeline corridor from Denver City to Clairemont, Texas: Lubbock, Museum of Texas Tech University, Lubbock Lake Landmark Quaternary Research Center Series 6, 238 p.

Johnson, E., and Holliday, V. T., 1980, A Plainview kill/butchering locale on the Llano Estacado—the Lubbock Lake site: Plains Anthropologist, v. 25, p. 89–111.

Johnson, E., and Holliday, V. T., 1986, The Archaic record at Lubbock Lake: Plains Anthropologist Memoir 21, p. 7–54.

Kelley, J. H., 1974, A brief resume of artifacts collected at the Lubbock Lake site prior to 1961, in Black, C. C., ed., History and prehistory of the Lubbock Lake site: Lubbock, West Texas Museum Association, The Museum Journal, v. 15, p. 15–42.

Kelley, V. C., 1972, Geology of the Fort Sumner sheet, New Mexico: New Mexico Bureau of Mines and Mineral Resources Bulletin 98, 55 p.

Kissinger, D. G., 1964, Circulionidae of America north of Mexico; a key to the genera: South Lancaster, Massachusetts, Taxonomic Publications, 143 p.

Klein, R. L., and Geis, J. W., 1978, Biogenic silica in the Pinaceae: Soil Science, v. 126, p. 145–155.

Knox, J. C., 1983, Responses of river systems to Holocene climates, in Wright, H. E., ed., Late-Quaternary environments of the United States, Vol. 2, the Holocene: Minneapolis, University of Minnesota Press, p. 26-41.

Koivo, L. K., 1976, Species diversity in postglacial diatom lake communities of Finland: Palaeogeography, Palaeoclimatology, Palaeoecology, v. 19, p. 165–190.

Krammer, K., and Lange-Bertalot, H., 1986, Susswasserflora von Mitteleuropa, Bacillariophyceae 1. Teil: Naviculadeae: Jena, Gustav Fischer Verlag, 876 p.

Krammer, K., and Lange-Bertalot, H., 1988, Susswasserflora von Mitteleuropa, Bacillariophyceae 2. Teil: Bacillariaceae, Epithemiaceae, Surirellaceae: Jena, Gustav Fischer Verlag, 596 p.

Krammer, K., and Lange-Bertalot, H., 1991, Susswasserflora von Mitteleuropa, Bacillariophyceae, Vol. 2, No. 3. Teil: Centrales, Fragilariaceae, Euno-

tiaceae: Jena, Gustav Fischer Verlag, 576 p.

Kutzbach, J. E., 1987, Model simulations of the climatic patterns during the deglaciation of North America, in Ruddiman, W. F., and Wright, H. E., Jr., eds., North America and adjacent oceans during the last deglaciation: Boulder, Colorado, Geological Society of America, The Geology of North America, v. K-3, p. 425–446.

Larkin, T. J., and Bomar, G. W., 1983, Climatic atlas of Texas: Austin, Texas Department of Water Resources LP-192, 151 p.

Lindroth, C. H., 1968, The ground beetles of Canada and Alaska, Part 5: Opuscula Entomologica Supplement No. 33, p. 649–944.

Lohman, K. E., 1936, Diatoms from Quaternary lake beds near Clovis, New Mexico: Journal of Paleontology, v. 9, p. 455–459.

Lotspeich, F. B., and Everhart, M. E., 1962, Climate and vegetation as soil forming factors on the Llano Estacado: Journal of Range Management, v. 15, p. 134–141.

Lowe, R. L., 1974, Environmental requirements and pollution tolerance of freshwater diatoms: Cincinnati, National Environmental Research Center, Office of Research and Development, U.S. Environmental Protection Agency, EPA-670/ 4-74-005, 333 p.

Lundelius, E. L., Jr., 1972, Vertebrate remains from the Gray Sand, in Hester, J. J., compiler, Blackwater Locality No. 1: A stratified Early Man site in eastern New Mexico: Taos, Publication of the Fort Burgwin Research Center, v. 8, p. 148–163.

Machette, M. N., 1985, Calcic soils of the southwestern United States: Geological Society of America Special Paper 203, p. 1–21.

Machette, M. N., 1986, Calcium and magnesium carbonates, in Singer, M. J., and Janitzky, P., eds., Field and laboratory procedures used in a soil chronosequence study: U.S. Geological Survey Bulletin 1648, p. 30–33.

Madole, R. F., Ferring, C. R., Guccione, M. J., Hall, S. A., Johnson, W. C., and Sorenson, C. J., 1991, Quaternary geology of the Osage Plains and Interior Highlands, in Morrison, R. B., ed., Quaternary nonglacial geology: Conterminous U.S.: Geological Society of America Centennial Volume K-2, p. 503–564.

Manabe, S., and Wetherald, R. T., 1986, Reduction in summer soil wetness induced by an increase in atmospheric carbon dioxide: Science, v. 232, p. 626–628.

Mandel, R. D., 1992, Geomorphology, in Saunders, J. W., Mueller-Wille, C. S., and Carlson, D. L., eds., An archaeological survey of the proposed South Bend Reservoir area: Young, Stephens, and Throckmorton Counties, Texas: Texas A&M University, Archeological Research Laboratory, Archeological Surveys 6, p. 53–83.

Matthews, J. A., 1980, Some problems and implications of ^{14}C dates from a podzol buried beneath an end moraine at Haugabreen, southern Norway: Geografiska Annaler, v. 62A, p. 185–208.

Matthews, J. A., 1985, Radiocarbon dating of surface and buried soils: principles, problems and prospects, in Richards, K. S., Arnett, R. R., and Ellis, S., eds., Geomorphology and soils: London, Allen and Unwin, p. 269–288.

McGrath, D. A., 1984, Morphological and mineralogical characteristics of indurated caliches of the Llano Estacado [M.S. thesis]: Lubbock, Texas Tech University, 206 p.

Meade, G. E., 1945, The Blanco fauna: The University of Texas Publication 4401, p. 509–556.

Meltzer, D. J., 1991, Altithermal archaeology and paleoecology at Mustang Springs, on the Southern High Plains of Texas: American Antiquity, v. 56, p. 236–267.

Meltzer, D. J., 1995, Modelling the prehistoric response to Altithermal climates on the Southern High Plains, in Johnson, E., ed., Ancient peoples and landscapes: Lubbock, Texas Tech University Press, p. 349–368.

Meltzer, D. J., and Collins, M. B., 1987, Prehistoric water wells on the Southern High Plains: clues to Altithermal climate: Journal of Field Archaeology, v. 14, p. 9–27.

Muhs, D. R., and Maat, P. B., 1993, The potential response of Great Plains eolian sands to greenhouse warming and precipitation reduction: Journal of Arid Environments, v. 25, p. 351–361.

Nativ, R., 1988, Hydrogeology and hydrochemistry of the Ogallala aquifer, Southern High Plains, Texas Panhandle and eastern New Mexico: The University of Texas at Austin Bureau of Economic Geology, Report of Investigations 177, 64 p.

Neck, R. W., 1987, Changing Holocene snail faunas and environments along the eastern Caprock escarpment of Texas: Quaternary Research, v. 27, p. 312–322.

Neck, R. W., 1994, Appendix E: Analysis of gastropod remains from prehistoric wells at Sulphur Springs Reservoir site 41MT21, *in* Quigg, J. M., Frederick, C. D., and Lintz, C., authors, Sulphur Springs Draw: Geoarchaeological and archaeological investigations at Sulphur Draw Reservoir, Martin County, Texas: Austin, Mariah Associates, Inc., Mariah Technical Report 776, p. E1–E4.

NOAA (National Oceanic and Atmospheric Administration), 1982, Climate of Texas: Asheville, North Carolina, National Climatic Data Center, Climatography of the United States, No. 60, 46 p.

Nordt, L. C., Boutton, T. W., Hallmark, C. T., and Waters, M. R., 1994, Late Quaternary vegetation and climate changes in central Texas based on the isotopic composition of organic carbon: Quaternary Research, v. 41, p. 109–120.

Norgren, J., 1973, Distribution, form and significance of plant opal in Oregon soils [Ph.D. thesis]: Corvallis, Oregon State University, 176 p.

Oldfield, F., 1975, Pollen-analytical results, Part II, *in* Wendorf, F., and Hester, J. J., eds., Late Pleistocene environments of the Southern High Plains: Taos, Fort Burgwin Research Center Publication 9, p. 121–147.

Oldfield, F., and Schoenwetter, J., 1964, Late Quaternary environments of Early Man on the Southern High Plains: Antiquity, v. 38, p. 226–229.

Parry, W. J., and Speth, J. D., 1984, The Garnsey Spring campsite: Late Prehistoric occupation in southeastern New Mexico: University of Michigan, Museum of Anthropology, Technical Report 10, 228 p.

Patrick, R., 1938, The occurrence of flints and extinct animals in pluvial deposits near Clovis, New Mexico, Part V: diatom evidence from the mammoth pit: Proceedings, Philadelphia Academy of Natural Science, v. 90, p,. 15–24.

Patrick, R., 1975, The diatoms of the United States exclusive of Alaska and Hawaii, Volume 2, Part 1: Academy of Natural Sciences of Philadelphia Monographs, Number 13, 213 p.

Patrick, R., and Reimer, C. W., 1966, The diatoms of the United States, Volume 1: Academy of Natural Sciences of Philadelphia Monographs, Number 13, 688 p.

Pendall, E., and Amundson, R., 1990, Stable isotope chemistry of pedogenic carbonate in an alluvial soil from the Punjab, Pakistan: Soil Science, v. 149, p. 199–211.

Pierce, H. G., 1974, The Blanco beds, *in* Meade, G., Evans, G. L., and Brand, J. P., eds., Guidebook to the Mesozoic and Cenozoic geology of the southern Llano Estacado: Lubbock, Texas, Lubbock Geological Society, p. 9–16.

Pierce, H. G., 1987, The gastropods, with notes on other invertebrates, *in* Johnson, E., ed., Lubbock Lake: late Quaternary studies on the Southern High Plains: College Station, Texas A&M University Press, p. 41–48.

Pilsbry, H. A., 1935, Report on shells collected by E. B. Howard from lake bed southwest of Clovis, Roosevelt County, New Mexico, *in* Howard, E. B., author, Evidence of Early Man in North America: University of Pennsylvania, The University Museum, The Museum Journal, v. 24, p. 89–90.

Pilsbry, H. A., 1939–1948, Land Mollusca of North America (north of Mexico): Philadelphia Academy of Natural Science Monographs 3, vol. I, p. 1–994; vol. II, p. 1–1113.

Piperno, D. R., 1988, Phytolith analysis—an archaeological and geological perspective: New York, Academic Press, 280 p.

Price, W. A., 1944, The Clovis site: regional physiography and geology: American Antiquity, v. 9, p. 401–407.

Quigg, J. M., Frederick, C. D., and Lintz, C., 1993, Archaeological and geomorphological investigations at Red Lake Dam axis, borrow area, and spillway, Martin County, Texas: Austin, Mariah Associates, Inc., Mariah Technical Report 873, 107 p.

Quigg, J. M., Frederick, C. D., and Lintz, C., 1994, Sulphur Springs Draw: Geoarchaeological and archaeological investigations at Sulphur Draw Reservoir, Martin County, Texas: Austin, Mariah Associates, Inc., Mariah Technical Report 776, 189 p.

Reeves, C. C., Jr., 1965, Chronology of west Texas pluvial lake dunes: Journal of Geology, v. 73, p. 504–508.

Reeves, C. C., Jr. 1966, Pluvial lake basins of west Texas: Journal of Geology, v. 74, p. 269–291.

Reeves, C. C., Jr., 1970, Drainage pattern analysis, Southern High Plains, West Texas and New Mexico, *in* Mattox, R. B., and Miller, W. D., eds., Ogallala aquifer symposium: Lubbock, Texas Tech University, ICASALS (International Center for Arid and Semi-Arid Land Studies), p. 58–71.

Reeves, C. C., Jr., 1972, Tertiary-Quaternary stratigraphy and geomorphology of west Texas and southeastern New Mexico, *in* Kelley, V., and Trauger, F. D., eds., Guidebook for east-central New Mexico: New Mexico Geological Society Guidebook 24, p. 108–117.

Reeves, C. C., Jr., 1976, Quaternary stratigraphy and geological history of the Southern High Plains, Texas and New Mexico, *in* Mahaney, W. C., ed., Quaternary stratigraphy of North America: Stroudsburg, Pennsylvania, Dowden, Hutchinson and Ross, Inc., p. 213–234.

Reeves, C. C., Jr., 1990, Structural and geomorphic effects of Permian salt dissolution near Anton, Texas: Lubbock, Texas Tech University, unpublished report, 20 p.

Round, F. E., 1981, The ecology of the algae: Cambridge, Cambridge University Press, 653 p.

Rovner, I., 1971, Potential of opal phytoliths for use in paleoecological reconstruction: Quaternary Research, v. 1, p. 343–359.

Russell, R., 1945, Climates of Texas: Annals of the Association of American Geographers, v. 35, p. 37–52.

Scharpenseel, H. W., 1971, Radiocarbon dating of soils: Soviet Soil Science, v. 3, p. 76–83.

Scharpenseel, H. W., 1979, Soil fraction dating, *in* Berger, R., and Suess, H., eds., Radiocarbon dating, Proceedings, International Radiocarbon Conference, 9th, Los Angeles and La Jolla, 1976: Berkeley, University of California Press, p. 277–283.

Schneider, S. H., 1989, The Greenhouse effect: science and policy: Science, v. 243, p. 771–781.

Schoenwetter, J., 1975, Pollen-analytical results, Part I, *in* Wendorf, F., and Hester, J. J., eds., Late Pleistocene environments of the Southern High Plains: Taos, Fort Burgwin Research Center Publication 9, p. 103–120.

Schultz, G. E., 1990a, Biostratigraphy and volcanic ash deposits of the Tule Formation, Briscoe County, Texas, *in* Gustavson, T. C., ed., Tertiary and Quaternary stratigraphy and vertebrate paleontology of parts of northwestern Texas and eastern New Mexico: The University of Texas at Austin Bureau of Economic Geology, Guidebook 24, p. 60–64.

Schultz, G. E., 1990b, Blanco local fauna and the Blancan Land Mammal age, *in* Gustavson, T. C., ed., Tertiary and Quaternary stratigraphy and vertebrate paleontology of parts of northwestern Texas and eastern New Mexico: The University of Texas at Austin Bureau of Economic Geology, Guidebook 24, p. 44–51.

Sellards, E. H., 1938, Artifacts associated with fossil elephant: Geological Society of America Bulletin, v. 49, p. 999–1010.

Sellards, E. H., 1952, Early Man in America: Austin, University of Texas Press, 211 p.

Sellards, E. H., 1955a, Fossil bison and associated artifacts from Milnesand, New Mexico: American Antiquity, v. 20, p. 336–344.

Sellards, E. H., 1955b, Further investigations at the Scharbauer site, *in* Wendorf, F., Krieger, A. D., Albritton, C. C., Jr., and Stewart, T. D., authors, The Midland discovery: Austin, University of Texas Press, p. 126–132.

Sellards, E. H., and Evans, G. L., 1960, The Paleo-Indian cultural succession in the central High Plains of Texas and New Mexico, *in* Wallace, A. F. C., ed., Men and cultures: Philadelphia, University of Pennsylvania Press,

p. 639–649.

Sellards, E. H., Evans, G. L., and Meade, G. E., 1947, Fossil bison and associated artifacts from Plainview, Texas: Geological Society of America Bulletin, v. 58, p. 927–954.

Singer, A., 1989, Palygorskite and sepiolite group minerals, in Dixon J. B., and Weed, S. B., eds., Minerals in soil environments (second edition): Madison, Wisconsin, Soil Science Society of America, p. 829–872.

Singer, M. J., 1986, Bulk density – paraffin clod method, in Singer, M. J., and Janitzky, P., eds., Field and laboratory procedures used in a soil chronosequence study: U.S. Geological Survey Bulletin 1649, p. 18–19.

Slaughter, B. H., 1975, Ecological interpretations of Brown Sand Wedge local fauna, in Wendorf, F., and Hester, J. J., eds., Late Pleistocene environments of the Southern High Plains: Taos, Fort Burgwin Research Center Publication 9, p. 179–192.

Smith C., Runyon, J., and Agogino, G., 1966, A progress report on a Pre-Ceramic site at Rattlesnake Draw, eastern New Mexico: Plains Anthropologist, v. 11, p. 302–313.

Smith, J. B., and Tirpak, D. A., 1990, The potential effects of global climate change on the United States: New York, Hemisphere Publishing Corporation, 689 p.

Soil Survey Division Staff, 1993, Soil survey manual: U.S. Department of Agriculture Handbook 18, 437 p.

Soil Survey Staff, 1990, Keys to soil taxonomy (fourth edition): SMSS Technical Monograph 19, Virginia Polytechnic Institute, 422 p.

Speer, R., 1990, History of the Plainview site, in Holliday, V. T., and Johnson, E., eds., Guidebook to the Quaternary history of the Llano Estacado: Lubbock, Texas Tech University, Lubbock Lake Landmark Quaternary Research Series, v. 2, p. 79–92.

Stafford, T. W., Jr., 1981, Alluvial geology and archaeological potential of the Texas Southern High Plains: American Antiquity, v. 46, p. 548–565.

Stafford, T. W., Jr., 1984, Quaternary stratigraphy, geochronology, and carbon isotope geology of alluvial deposits in the Texas Panhandle [Ph.D. thesis]: Tucson, The University of Arizona, 161 p.

Stanford, D., Haynes, C. V., Jr., Saunders, J. J., Agogino, G. A., and Boldurian, A. T., 1990, Blackwater Draw Locality 1: History, current research, and interpretations, in Holliday, V. T., and Johnson, E., eds. Guidebook to the Quaternary history of the Llano Estacado: Lubbock, Texas Tech University, Lubbock Lake Landmark Quaternary Research Series, v. 2, p. 105–155.

Stevens, D., 1973, Blackwater Draw Locality No. 1, 1963–1972, and its relevance to the Firstview Complex [M.A. thesis]: Portales, Eastern New Mexico University, 90 p.

Stock, C., and Bode, F. D., 1936, The occurrence of flints and extinct animals in pluvial deposits near Clovis, New Mexico, Part III: geology and vertebrate paleontology of the Quaternary near Clovis, New Mexico: Proceedings, Philadelphia Academy of Natural Sciences, v. 88, p. 219–241.

Stuiver, M., 1993, Editorial comment: Radiocarbon, v. 35, p. iii.

Stuiver, M., and Pearson, G. W., 1992, Calibration of the radiocarbon time scale, 2500–5000 BC, in Taylor, R. E., Long, A. and Kra, R. S., eds., Radiocarbon after four decades: an interdisciplinary perspective: New York, Springer-Verlag, p. 19–33.

Tamayo, J. L., 1962, Geografica general de Mexico: Mexico City, Instituto Mexicano de Investigaciones Economicas, 22 p.

Tanner, C. B., and Jackson, M. L., 1947, Nomographs of sedimentation times for soil particles under gravity or centrifugal acceleration: Soil Science Society of America Proceedings, v. 12, p. 60–65.

Tempere, J. A., and Peragallo, H., 1907–1915, Diatomees du Monde Entier, edition 2, 30 fasc.: Arcachon, Grez-sur-Loing (S.-et-M.), fascicule 1, p. 1–16, 1907; 2–7, p. 17–112, 1908; 8–12, p. 113–208, 1909; 13–16, p. 209–256, 1910; 17–19, 257–304, 1911; 20–23, p. 305–352, 1912; 24–28, p. 353–448, 1913; 29–30, p. 449–480, 1914; Tables, p. 1–68, 1915.

Theis, C. V., 1932, Report on the ground water in Curry and Roosevelt Counties, New Mexico: New Mexico State Engineer's Office, 10th Biennial Report, p. 98–160.

Thompson, J. L., 1987, Modern, historic, and fossil flora, in Johnson, E., ed., Lubbock Lake: late Quaternary studies on the Southern High Plains: College Station, Texas A&M University Press, p. 26–35.

Tieszen, L. L., 1994, Stable isotopes on the Great Plains: Vegetation analyses and diet determinations, in Owsley, D., and Jantz, R., eds., Skeletal biology and the Great Plains: Washington, D.C., Smithsonian Institution Press, p. 261–282.

Tomanek, G. W., and Hulett, G. K., 1970, Effects of historical droughts on grassland vegetation in the Central Great Plains, in Dort, W., Jr., and Jones, J. K., Jr., eds., Pleistocene and Recent environments of the central Great Plains: Lawrence, University of Kansas Press, p. 203–210.

Twiss, P. C. 1987, Grass-opal phytoliths as climatic indicators of the Great Plains Pleistocene, in Johnson, W. C., ed., Quaternary environments of Kansas: Kansas Geological Guide Book, Series 5, p. 179–188.

Waker, J. R., 1978, Geomorphic evolution of the Southern High Plains: Baylor Geological Studies Bulletin 35, 32 p.

Weaver, J. E., and Albertson, F. W., 1943, Resurvey of grasses, forbs, and underground plant parts at the end of the great drought: Ecological Monographs, v. 13, p. 64–117.

Webb, T., III, Ruddiman, W. F., Street-Perrott, V. A., Markgraf, V., Kutzbach, J. E., Bartlein, P. J., Wright, H. E., Jr., and Prell, W. L., 1993, Climatic changes during the past 18,000 years: Regional syntheses, mechanisms, and causes, in Wright, H. E., Jr., Kutzbach, J. E., Webb, T., III, Ruddiman, W. F., Street-Perrott, F. A., and Bartlein, P. J., eds., Global climates since the last glacial maximum: Minneapolis, University of Minnesota Press, p. 514–535.

Weeks, J. B., and Gutentag, E. D., 1988, Region 17, High Plains, in Back, W., Rosenshein, J. S., and Seaber, P. R., eds., Hydrogeology: Boulder, Geological Society of America, The Geology of North America, v. 0-2, p. 157–164

Wendorf, F., assembler, 1961a, Paleoecology of the Llano Estacado: Santa Fe, The Museum of New Mexico Press, Fort Burgwin Research Center Publication 1, 144 p.

Wendorf, F., 1961b, An interpretation of late Pleistocene environments of the Llano Estacado, in Wendorf, F., assembler, Paleoecology of the Llano Estacado: Santa Fe, The Museum of New Mexico Press, Fort Burgwin Research Center Publication 1, p. 115–133.

Wendorf, F., 1970, The Lubbock Subpluvial, in Dort, W., and Jones, J. K., eds., Pleistocene and Recent environments of the Central Great Plains: Lawrence, The University of Kansas Press, p. 23–36.

Wendorf, F., 1975a, The modern environment, in Wendorf, F., and Hester, J. J., eds., Late Pleistocene environments of the Southern High Plains: Taos, Fort Burgwin Research Center Publication 9, p. 1–12.

Wendorf, F., 1975b, Summary and conclusions, in Wendorf, F., and Hester, J. J., eds., Late Pleistocene environments of the Southern High Plains: Taos, Fort Burgwin Research Center Publication 9, p. 257–278.

Wendorf, F., and Hester, J. J., 1962, Early Man's utilization of the Great Plains environment: American Antiquity, v. 28, p. 159–171.

Wendorf, F., and Hester, J. J., eds., 1975, Late Pleistocene environments of the Southern High Plains: Taos, Fort Burgwin Research Center Publication 9, 290 p.

Wendorf, F., and Krieger, A. D., 1959, New light on the Midland discovery: American Antiquity, v. 25, p. 66–78.

Wendorf, F., Krieger, A. D., Albritton, C. C., Jr., and Stewart, T. D., 1955, The Midland discovery: Austin, University of Texas Press, 139 p.

Wheat, J. B., 1974, First excavations at the Lubbock Lake site, in Black, C. C., ed., History and prehistory of the Lubbock Lake site: West Texas Museum Association, The Museum Journal, v. 15, p. 15–42.

White, W. N., Broadhurst, W. L., and Lang, J. W., 1946, Ground water in the High Plains of Texas: U.S. Geological Survey Water-Supply Paper 889-F, p. 381–420.

Wilding, L. P., and Drees, L. R., 1974, Contributions of forest opal and associated crystalline phases of fine clay fractions of soils: Clays and Clay

Minerals, v. 22, p. 295–306.

Wilding, L. P., Smeck, N. E., and Drees, L. R., 1977, Silica in soils: quartz, cristobalite, tridymite, and opal, *in* Dixon, J. B., and Weed, S. B., eds., Minerals in soil environments: Madison, Wisconsin, Soil Science Society of America, p. 471–552.

Willey, P. S., and Hughes, J. T., 1978, The Deadman's shelter site, *in* Hughes, J. T., and Willey, P. S., eds., Archaeology at MacKenzie Reservoir: Austin, Texas Historical Commission Archeological Survey Report, v. 24, p. 149–204.

Winkler, D. A., 1987, Vertebrate-bearing eolian unit from the Ogallala Group (Miocene) in northwestern Texas: Geology, v. 15, p. 705–708.

Winsborough, B. M., 1988, Paleoecological analysis of Holocene algal mat diatomites associated with prehistoric wells on the Texas High Plains: Geological Society of America Abstracts with Programs, v. 20, p. 132.

Winborough, B. M., 1990, Some ecological aspects of modern fresh-water stromatolites in lakes and streams of the Cuatro Ciénegas Basin, Coahuila, Mexico [Ph.D. thesis]: Austin, The University of Texas, 341 p.

MANUSCRIPT ACCEPTED BY THE SOCIETY NOVEMBER 23, 1994

Index

[Italic page numbers indicate major references]

A

Acanthoscelides, 52
Aegialia, 51
Agonum, 51
algae
  golden, 48
  siliceous bodies, 47, 48, 50
algal fossils, 67
alluvial sand, 1
alluvium, 29
Amara, 51
Anametis subfusca, 51
Anderson Basin site
  molluscs, 65, 66
  stratum 1, 62
  stratum 2, 35, 44, 85
Anderson Basin #1, 43
  diatoms, 84
  molluscs, 42
  stratum 1, 42
  stratum 2, 43
  stratum 3, 37, 43
  stratum 4, 38
Anderson Basin #2
  diatoms, 84
  stratum 1, 31, 32
  stratum 4, 38
Anomoeonsis spp., 67
Anton basin, 13, 14
Apache soil, 42
Aphodius, 51, 52
  crenicollis, 52
Aplexa hypnorum, 62
archaeological features, 32, 43, 91
Arch Lake basin. *See* Salt Lake
  basin
aridity, 2
arrowhead, 51
Artemisia spp., 52, 54
artifacts
  Clovis culture, 32
  Folsom culture, 32
  Paleoindian, 2, 36, 44
Atriplex spp., 52

B

Bailey Draw, 10, 54
Baker site, 37, 40
Barwise, 39
beetles, fossil, 47, 50, 51, 52
Ben site, 26
BFI site, stratum 3, 37
biosilicate assemblages, 49
Birdwell site, 40
Bison, 55, 56, 57
  antiquus, 32, 36, 88
bison kills, bone beds, 43
bivalves, 46, 47
Blackwater Draw, 1, 5, 6, 10, 13, 15,
    18, 19, 20, 22, 24, 41, 45, 85, 93,
    94

Blackwater Draw (continued)
  ancestral, 18
  diatoms, 67
  drainage patterns, 25
  fossils, 47
  geochronology, 86
  lower, 23, 26, 85
    meanders, 29
  meanders, 25
  middle, 26
    springs, 29
    stratum 4, 38, 89
  molluscan shells, 59
  ostracodes, 59
  paleolakes, 82
  slopes, 23
  springs, 92
  stratum 1, 31
  stratum 2, 36
  stratum 3, 37
  stratum 5, 41, 42
  upper
    stratum 2, 44, 85
    stratum 4, 89
  Blackwater Draw Formation, 11, 13,
    15, 19, 20, 22, 23, 24, 35, 37, 45
Blackwater system, 24
Blackwater/Yellowhouse Canyon
    drainage, 17
Blancan Land Mammal age, 13
Blanco basin, 13
Blanco Canyon, 18, 39
Blanco Formation, 11, 13, 15, 35
Bledsoe, 38
Bouteloua-Buchloe, 48
Bouteloua sp., 5
Bovina site, 94
Brazos River, 18
  ancestral, 17, 24
  drainages, 84, 85
  headwaters, 29
  tributaries, 1, 4, 10
Brazos River system, 8, 13, 15, 26, 62,
    82
  ancient, 16
  draws, 22, 37, 91
  stratigraphic record, 82
Brooks site, 48
Brownfield Lake, 29, 31
Brownfield site, 31, 59
Bruchidae, 52
Buchloe dactyloides, 5
buffalo grass, 5
bulrush, 48, 49, 50

C

calcium, 59
calcium bicarbonate, 59
calcium carbonate, 59, 94
calcrete
  Blanco Formation, 11

calcrete (continued)
  Caprock, 35
  lacustrine, 32
  silicified, stratum 1, 31
Calosoma porosifrons, 51
Canadian River valley, 4, 91
Candona spp., 59
Caprock calcrete, 35
Caprock escarpment, 66, 92
Carabidae, 51
Car Body locality, 44
  carbonate, 29, 94
  lacustrine, 1
  paludal, 29
Carex spp., 51
Chara sp., 59
Chlamydotheca spp., 59
chloride, 59
Chloridoid, 48, 49, 50
Citellus, 52
cliffrose, 51
climate, 4, 52, 56, 57, 73, 89, 90, 91
  changes, 2, 92, 93
  trends, 91
Clovis culture, 88
  artifacts, 32
Clovis gravel pit, 31, 44
Clovis paleobasin, 44
Clovis site, 2, 5, 6, 8, 10, 19, 20, 26,
    32, 39, 40, 67, 86, 94, 120
  diatoms, 67, 73, 84
  drought, 90
  dunes, 85
  eolian deposits, 89, 91
  molluscs, 63, 65, 82
  occupation, 92
  paleolakes, 82
  pollen, 53, 54
  springs, 29, 92
  stratum 1, 88
  stratum 2, 35, 37, 44, 85
  stratum 4, 38
  stratum 5, 42
Cocconeis placentula, 68
Colorado Plateau, meanders, 25
Colorado River, 20
  ancestral, 21
  drainages, 37, 84, 85
  tributaries, 1, 4, 10
Colorado River system, 8, 13, 15, 26,
    62, 85
  draws, 22, 23, 91
  stratigraphic record, 82
Columella columella, 65
County Caliche Pit, 41
Cowania ericaefolia, 51
Coyote Lake area, 17
Coyote Lake basin, 20
Cratacanthus dubius, 52
creosote bush, 52
Cuprideis salebrosa, 59
Curculionidae, 51, 52

Typeset in U.S.A. by Johnson Printing, Boulder, Colorado
Printed in U.S.A. by Malloy Lithographing, Inc., Ann Arbor, Michigan